Featherstone

Jenny Aitken, Jan Hunt, Elizabeth Roy, Bess Sajfar and Sally Featherstone

A SENSE OF WONDER

Science in early childhood education

Featherstone Education
An imprint of Bloomsbury Publishing Plc

50 Bedford Square   1385 Broadway
London   New York
WC1B 3DP   NY 10018
UK   USA

www.bloomsbury.com

Bloomsbury is a registered trademark of Bloomsbury Publishing Plc

Text © Jenny Aitken, Jan Hunt, Elizabeth Roy, Bess Sajfar and Sally Featherstone, 2015
This UK edition of *A Sense of Wonder* is published by Bloomsbury Publishing Plc
by arrangement with Teaching Solutions, Australia
Interior photographs © Teaching Solutions/© Shutterstock/ © LEYF/ © Acorn Nursery, 2015
Cover photographs © London Early Years Foundation, 2015
Design by Lynda Murray, 2015

British Library Cataloguing-in-Publication Data
A catalogue record for this book is available from the British Library.

ISBN
Paperback 978-1-4729-1340-1
ePDF 978-1-4729-1341-8

Library of Congress Cataloging-in-Publication Data
A catalogue record for this book is available from the Library of Congress.

1 3 5 7 9 10 8 6 4 2

Printed and bound in India by Replika Press Pvt. Ltd.

This book is produced using paper that is made from wood grown in managed, sustainable
forests. It is natural, renewable and recyclable. The logging and manufacturing processes
conform to the environmental regulations of the country of origin.

To view more of our titles please visit www.bloomsbury.com

Foreword

As science teachers in both higher education and Further Education (FE), we enjoy a shared passion for play-based science learning and the provision of meaningful and inviting opportunities for young children.

The idea for this book arose from our commitment to providing annual science conferences for early childhood student practitioners. Our students would go away and incorporate the ideas we discussed on placement and a year later would return as inspired science practitioners, glowing with stories of children's wonder and curiosity about the simple principles of science in the world around them. We are appreciative of their anecdotes as they have inspired us to write this book.

We hope our book will invite practitioners to tap into and extend on the innate sense of wonder that children possess as they delight in exploring, discovering and experimenting in the natural world. Each chapter includes links to the Early Years Foundation Stage to empower practitioners to confidently implement science with all age groups in early childhood education.

Many practitioners already involve children in rich learning experiences, so this book will validate existing practice; however, it is hoped that the ideas presented will inspire further science learning. When practitioners embark on a journey of science with children there can be a profound impact on each child's future interests and learning of science.

Every practitioner can 'do' science!

Jenny Aitken, Jan Hunt, Elizabeth Roy and Bess Sajfar

The inspiration for this book

The authors of the original text drew inspiration from the Australian Early Years Learning Framework. This edition has adaptations of the original links to include reference to the EYFS (England) and is coherent with the guidance for the Early Years in Scotland, Wales and Northern Ireland.

Efforts have been made to link scientific language and processes to help practitioners to put the EYFS, or the local requirements into practice in a scientific context. References and relevant quotes help practitioners to make useful connections between the areas of learning and their local curriculum guidance. The book will also ensure the formation of a strong and relevant foundation for children to continue building their learning about science in primary school.

Where we make reference to detailed assessment criteria in the curriculum for science, we have used the Early Years Foundation Stage (England).

During the first stages of scientific learning, young children are seen as confident, capable, involved learners, eager to observe and explore the world around them. This can be supported by creating an active, enquiry-based approach within a play-based programme, and promoting positive dispositions for learning in all areas of the curriculum.

As they experience and get involved in scientific activities, children develop a range of skills and engage in a range of scientific processes. As identified in the EYFS, these processes include problem-solving, inquiring, experimenting, hypothesising, researching and investigating.

Sally Featherstone

Contents

Chapter 1

What is science?

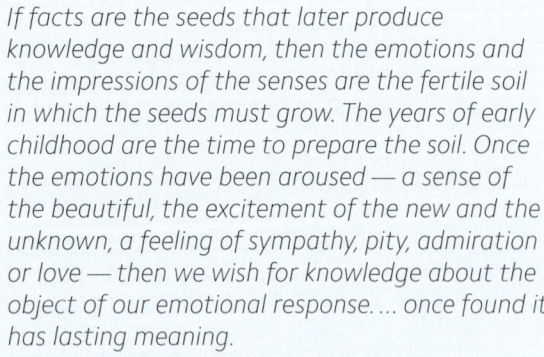

> If facts are the seeds that later produce knowledge and wisdom, then the emotions and the impressions of the senses are the fertile soil in which the seeds must grow. The years of early childhood are the time to prepare the soil. Once the emotions have been aroused — a sense of the beautiful, the excitement of the new and the unknown, a feeling of sympathy, pity, admiration or love — then we wish for knowledge about the object of our emotional response.... once found it has lasting meaning.
>
> Rachel Carson

What is the place of science in the early years?

Science is all around us in our everyday lives. Put very simply, science is the study of our natural world.

In their document 'Where Does Science fit in the EYFS?', the Association for Science Education (ASE) describes how:

> *Science provides foundations for people to understand what is happening around them, supplies information for developments to be made and explain why things happen.*
> *'Science, as a subject, is intrinsically fascinating to children and involves them in exploration and 'finding out' for themselves'. (Farmery, 2002)*
>
> *Early Years: Where does science fit in?* ASE; 2012

In early years settings, enthusiastic practitioners encourage children to engage with the natural world and build scientific concepts, understandings and language.

This baby is conducting research too. He is using his senses to explore the properties of a lemon in his search for new scientific understanding about the natural world.

What is science?

> *Science is an intellectual activity … designed to discover information about the natural world in which humans live and to discover the ways in which this information can be organised into meaningful patterns.*
>
> Sheldon Gotleib; University of South Alabama

> *'Science' comes from the Latin word 'scientia', meaning knowledge.*
>
> Oxford English Dictionary

> *Science may be a particularly important domain in early years, serving not only to build a basis for future scientific understanding but also to build important skills and attitudes for learning.*
>
> Science in Early Childhood Classrooms: Content and Process; Karen Worth

Science in the Early Years Foundation Stage

Every practitioner can do science!

Through this book we hope to inspire early years practitioners to 'have a go' at science. Albert Einstein said 'science is having fun with ideas', and the science ideas in each chapter of this book, with their simple scientific explanations, help convey the principle that the science curriculum is easy to implement in early years settings.

In many instances, this book will validate what practitioners are already incorporating in their practice. We hope that the finer detail we have provided will enhance these daily activities for children and will also begin a wonderfully exciting journey for practitioners.

Where does science learning fit into an early years curriculum?

In early years settings, science can occur indoors and outdoors... in the sand... in the block area... in the home corner... in the vegetable patch... on a swing... coming down a slide... when caring for a pet... watching rain...

> *Science involves more than the gaining of knowledge. It is the systematic and organised enquiry into the natural world and its phenomena.*
>
> Vanderbilt University, Tennessee

A scientific exploration can start with the excitement of discovering a spider's web glistening with morning dew, which might lead to the creation of another spider's web made from twigs and wool (below).

This two-year-old is organising her understanding of reflections through conducting research into the properties of the mirror.

Through the provision of a play-based curriculum, practitioners are able to incorporate meaningful science experiences into their existing programmes. The photograph below shows how children are able to investigate the growing cycle of seedlings, which have been placed at their level, while the set-up of the environment invites participation, inquiry, research and exploration by the provision of magnifying glasses and drawing materials.

> *By the time children reach the age of five, they should know about similarities and differences in relation to places, objects, materials and living things. They talk about the features of their own immediate environment and how environments might vary from one another. They make observations of animals and plants and explain why some things occur, and talk about changes.*
>
> EYFS Profile, 2014, Department for Education

Investigating the growing cycle of seedlings.

As practitioners, we may be inspiring the world's future scientists. This scientist is conducting research into a drug that could control how energy is used within cells. This may help people who are obese and diabetic.

Scientific learning areas

When researching the science curriculum, you may encounter different names and grouping for the areas of science. For example, 'biological science' might be called 'life science', and 'earth science' might be called 'nature science'. Regardless of the terminology used, it is important to recognise the scientific learning opportunities supporting children to use correct scientific terminology and explore relevant scientific concepts.

In the EYFS curriculum for most UK countries, scientific learning is incorporated into an area that looks at the world, and includes other elements such as ICT, history, geography. In Northern Ireland the area is called '**The World Around Us**'; in Wales it is '**Knowledge and Understanding of the World**'; and in England it is '**Understanding the World**'. Scotland is the only country within the UK that separates early science as a discrete area of learning, called '**Sciences**'. In each country there is also guidance on active learning, thinking skills and the use of outdoor provision for scientific enquiry (see references, p.95).

The Australian Curriculum Framework for primary schools identifies the following science learning areas: biological science, physical science, chemical science, and earth and space science. These areas closely match the areas identified in the UK curriculum guidance for Key Stage 1, so we have retained this structure in this UK edition of this book.

- **biological science**

- **physical science** (including *chemical science*)

- **earth science** (*there is no focus on* **space science** *in this book because it is such an abstract concept for children*)

- **environmental science** (*we have included this because we believe sustainability and environmental education should underpin all science learning for young children*)

What is biological science?

Biological science is the study of living things.

Chapter five explores biological science and how it can be investigated in early childhood settings with a focus on:

- plants
- the importance of gardening
- animals
- the human body

Key question: What is alive?

Exploration of biological science in early childhood can include:

- plants and animals are living things
- living things have certain characteristics
- living things have parts that do different things
- living things change as they grow
- living things have basic needs
- living things have lifecycles
- living things live in a variety of habitats
- plants need food, sun and water
- plants grow from seeds
- animals need food and water
- some animals eat plants
- some animals eat other animals
- animals have babies
- my body is a living thing
- my body has parts that do different things

Growing plants is a practical way of investigating biological science.

What is physical science?

Physical science is the study of materials and energy in the non-living world.

Chapter six explores physical science and how it can be investigated in early childhood settings with a focus on:

- the nature of materials
- physical and chemical changes
- forces and movement of objects
- energy

Key question: How does it move?

Exploration of physical science in early childhood can include:

- some things melt when heated
- water is hard when it freezes
- food changes when it is cooked
- some things go rusty
- burning wood makes charcoal
- mixing colours makes new colours
- objects fall when dropped
- magnets push and pull each other
- some things float and others sink
- sound is made in lots of ways
- we can see reflections

What is earth science?

Earth science is the study of the earth and its materials.

Chapter seven explores earth science and how it can be investigated in early childhood settings with a focus on:

- non-living earth materials such as water, soil, rocks, sand and mud
- day and night
- weather and seasons

Key question: What are non-living things?

Exploration of earth science in early childhood can include:

- rain comes from clouds
- wind makes things move
- weather can change every day
- earth's materials have lots of uses
- in the day we see the sun in the sky
- we can see shadows on a sunny day
- water and soil make mud
- rocks can be different shapes and sizes

What is environmental science?

Environmental science is the study of caring for the natural world.

Chapter eight explores environmental science and how it can be investigated in early childhood settings with a focus on:

- composting
- worm farming
- environmentally sensitive early childhood programs
- using recycled materials

Key question: How can we care for our world?

Exploration of environmental science in early childhood can include:

- we can re-use things
- food scraps can be used as worm food
- worms can make food scraps turn into compost
- compost and worm manure help plants to grow
- waste can be managed
- plants need our care
- animals need our care
- weeding helps to give plants room to grow
- some plants can be eaten as food

Process skills used in science

In our society, scientists are involved in exploratory processes which aim to bring about new learning. This is known as research. Scientific research usually begins with a problem to be solved, followed by a systematic process of enquiry, where the problem is explored and further investigated, often by asking and answering questions.

Young children use a range of scientific process skills such as observing, questioning, predicting, comparing, identifying, measuring, classifying and communicating. It is exciting that these process skills form the basis of play-based scientific learning.

The six main process skills for science are outlined on these two pages.

Problem solving

Children...

- use play to solve problems and transfer knowledge from one situation to another

- apply thinking strategies and solve problems in new situations

- apply problem-solving strategies from one situation to a new context.

Enquiring

Children...

- wonder about what they see

- take part in scientific enquiry during discussions

- create and design ways to find answers to their questions.

Experimenting

Children...

- experiment , exploring cause and effect, often through trial and error

- mirror, repeat and practice actions of others immediately or later in their play.

This young child is applying problem-solving about force and motion as he pushes himself on the swing.

Experimenting with different ways of holding the spoon enables this young child to gain a sense of control over the process as she explores the properties of cake mix.

Hypothesising

Children...

- use reflective thinking to consider why things happen and what can be learned from these experiences

- make predictions and generalisations, and communicate this using scientific language

- make scientific connections between experiences, concepts and processes.

Researching

Children...

- use resources such as tools, everyday objects, books, media and ICT to find out more about their area of interest.

Investigating

Children...

- investigate, imagine and explore ideas through play

- use their senses to explore nature and the properties of objects.

> *To a child who's discovered the hammer, the whole world's a nail.*
>
> Jurgen Habermas

This saying illustrates a child's innate need to try things for himself. As practitioners, we reward this need to investigate by providing and guiding relevant scientific learning opportunities.

Young children can learn about reflections and shadows in water while observing objects floating and sinking.

Feeding the chickens, finding an egg, feeling the egg's warmth, wondering about how the chicken made it – this hands-on science learning involves making predictions, and forming biological scientific connections.

Chapter 2

Connecting science to the Early Years Foundation Stage

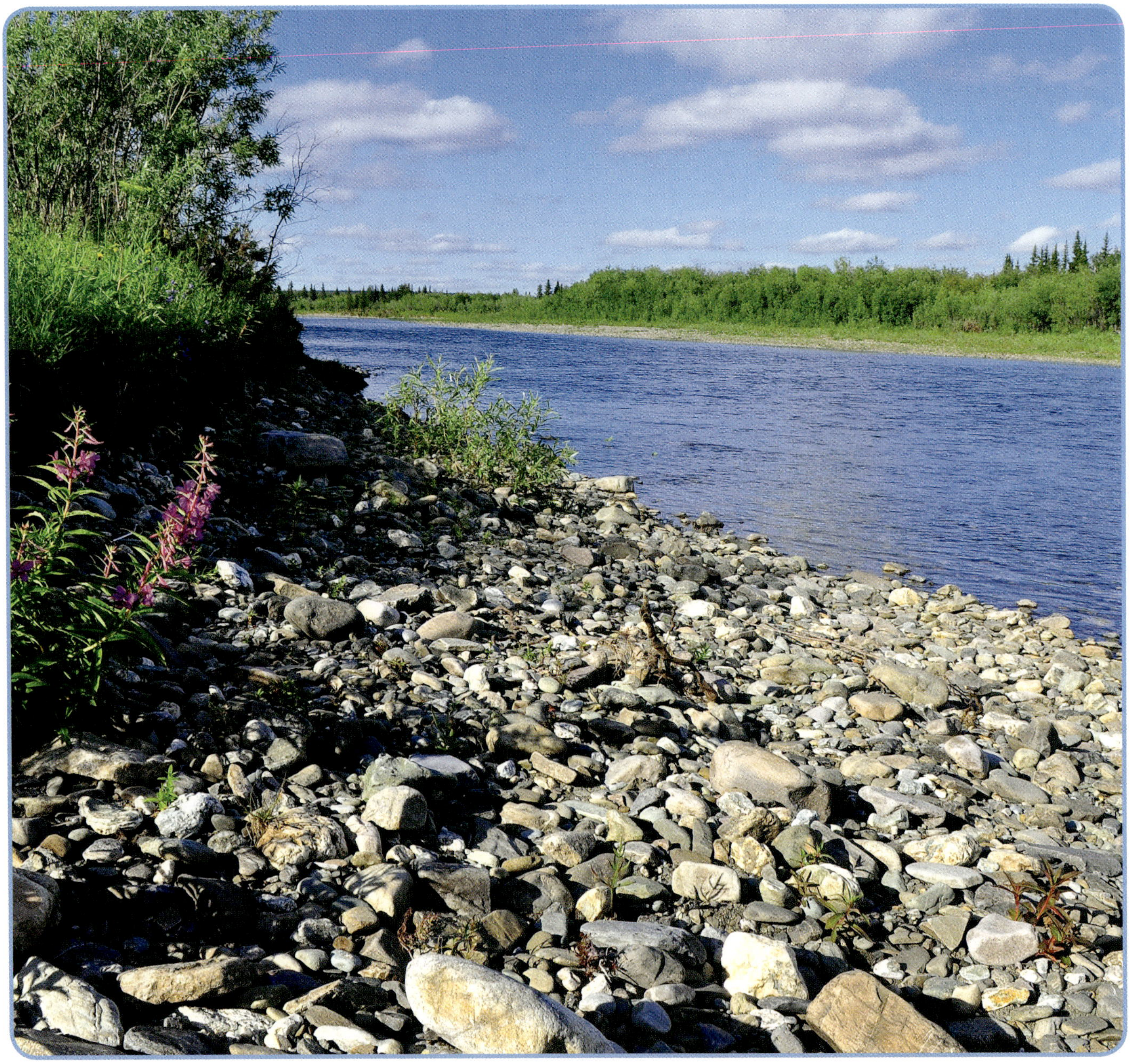

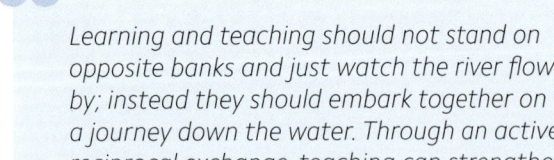

> *Learning and teaching should not stand on opposite banks and just watch the river flow by; instead they should embark together on a journey down the water. Through an active, reciprocal exchange, teaching can strengthen how to learn.*
>
> **Loris Malaguzzi**

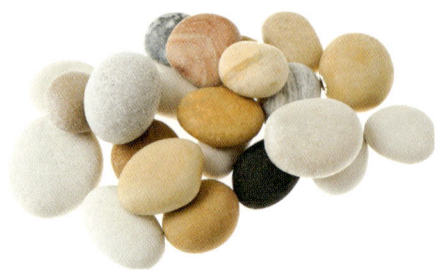

The EYFS framework

For ease of understanding, in the following sections, the Early Years Foundation Stage (EYFS) statutory framework and Development Matters guidance for England have been used as exemplars. References to the frameworks and guidance documents for other countries in the United Kingdom have been included as amplification.

The EYFS sets the standards that all early years providers must meet to ensure that children learn and develop well and are kept healthy and safe. The EYFS statute and guidance have been produced by the UK government to support the education and care of children from birth to the age of five.

> *This framework is mandatory for all early years providers (from 1 September 2012): maintained schools, non-maintained schools, independent schools, and all providers on the Early Years Register. The Early Years Foundation Stage (EYFS) sets the standards that all early years providers must meet to ensure that children learn and develop well and are kept healthy and safe. It promotes teaching and learning to ensure children's school readiness and gives children the broad range of knowledge and skills that provide the right foundation for good future progress through school and life.*
>
> **Statutory Framework for the Early Years Foundation Stage (England); 2014, Department for Education**

The national EYFS curriculum for England (2014), and Development Matters (British Association for early Childhood Education, 2012) provide guidance for early years practitioners to use when developing a curriculum for 0–5 year-old children. This book helps to link the EYFS principles, practice and learning outcomes to science learning.

The EYFS provides a broad direction for practitioners, teachers and their managers, who then need to think about their principles and practice when implementing the curriculum.

The Early Learning Framework for the Australian curriculum gives us a useful definition for the curriculum, which sits well with the curriculum guidance in the UK, and it is important that all practitioners understand that the curriculum covers every child's learning experience:

> *Curriculum — All the interactions, experiences, activities, routines and events, planned and unplanned, that occur in an environment designed to foster children's learning and development.*
>
> **Early Years Learning Framework EYLF (Australia); 2009, Department of Education, Employment and Workplace**

The themes, principles and practices of the EYFS

Every child is a **unique child**, who is constantly learning and can be resilient, capable, confident and self-assured

Children learn to be strong and independent through **positive relationships**

The **overarching principles** of the EYFS

Children learn and develop well in **enabling environments**, in which their experiences respond to their individual needs and there is a strong partnership between practitioners and parents and/or carers

Children develop and learn in different ways and at different rates. The framework covers the education and care of all children in early years provision, including children with special educational needs and disabilities

These four principles are founded on early years theories and research evidence of how children learn. They are the foundation upon which practitioners build their early years pedagogy. These principles also underpin the curriculum in Wales, Scotland and Northern Ireland.

> *Pedagogy — Central to the Foundation Phase approach is the practitioner as a facilitator of learning, with the child at the heart of learning and teaching. Care should be taken to ensure that the teaching approach used is based on the needs of the child in each area of learning and development.*
>
> **Learning and Teaching – Pedagogy; Welsh Assembly Government**

Practitioners' images of children and their beliefs about what children are capable of achieving will drive the way they interact with children and the type of science programme they deliver. If practitioners believe that children are strong and capable learners then they will be more likely to provide a challenging programme that encourages children to enquire and learn.

The themes and principles in the following table help practitioners to implement science learning in a positive, nurturing and stimulating learning environment.

Themes	A unique child	Positive relationships	Enabling environments	Learning and development
Principles	Every child is a unique child who is constantly learning and can be resilient, capable, confident and self-assured.	Children learn to be strong and independent through positive relationships.	Children learn and develop well in enabling environments, in which their experiences respond to their individual needs and there is a strong partnership between practitioners and parents and carers.	Children learn and develop in different ways. The framework covers the education and care of all children in early years provision, including children with special educational needs and disabilities. Practitioners teach children by ensuring challenging, playful opportunities across the prime and specific areas of learning and development.
Practice	Practitioners: ★ understand and observe each child's development and learning, assess progress, plan for next steps ★ support babies and children to develop a positive sense of their own identity and culture ★ identify any need for additional support ★ keep children safe ★ value and respect all children and families equally	Positive relationships are: ★ warm and loving, and foster a sense of belonging ★ sensitive and responsive to the child's needs, feelings and interests ★ supportive of the child's own efforts and independence ★ consistent in setting clear boundaries ★ stimulating ★ built on key person relationships in early years settings	Enabling environments: ★ value all people ★ value learning They offer: ★ stimulating resources, relevant to all the children's cultures and communities ★ rich learning opportunities through play and playful teaching ★ support for children to take risks and explore in a safe and controlled setting	They foster the characteristics of effective learning: ★ playing and exploring ★ active learning ★ creating and thinking critically

+   +   =

Development Matters in the EYFS © Crown copyright 2012

The early learning goals

The outcomes for the early years curriculum in England (EYFS) are clearly defined and include all areas of learning:

Communication and language
- Listening and attention
- Understanding
- Speaking

Physical development
- Moving and handling
- Health and self-care

Personal, social and emotional development
- Making relationships
- Self-confidence and self-awareness
- Managing feelings and behaviour

THE PRIME AREAS

THE SPECIFIC AREAS

Literacy
- Reading
- Writing

Mathematics
- Numbers
- Shape, space and measure

Understanding the world
- People and communities
- The world
- Technology

Expressive arts and design
- Exploring and using media and materials
- Being imaginative

EYFS Statutory Framework, DfE, 2014 © Crown copyright 2014

The learning that takes place depends on each child's abilities, disposition and learning preferences. When planning for scientific activities, practitioners should direct a significant amount of their focus on the relevant parts of Understanding the world:

> *The world: children know about similarities and differences in relation to places, objects, materials and living things. They talk about the features of their own immediate environment and how environments might vary from one another. They make observations of animals and plants and explain why some things occur, and talk about changes.*
>
> EYFS Statutory Framework, DfE, 2014

Of course, science does not exist in a separate 'box' in the curriculum. It will almost always appear in conjunction with evidence in other areas of learning, particularly in communication and language, and personal, social and emotional development. When referring to the learning outcomes it is important to make sure you record the specific learning outcomes that clearly describe what the child has achieved, even when your observation may take you into several areas of learning.

Children have a strong sense of identity when they:

✓ learn about themselves

✓ develop a sense of belonging as they participate in scientific experiences

✓ develop autonomy as they decide ways of exploring a science project

✓ develop confident self-identities as practitioners recognise and respect their contributions to the exploration of science

✓ see familiar items from home used in science explorations

✓ interact with others on a group science project

Harvesting herbs from the vegetable garden helps to develop a respect for the natural environment.

Children express interest and wonder in gardening in their own vegetable patch. Links are made to PSED (self-confidence and self-awareness), Understanding the world (the world), and Physical development (moving and handling).

Recognising that the person in the mirror is 'me' helps to build a sense of identity and an understanding of reflection. Links are made to PSED (self-confidence and self-awareness) and Understanding the world (the world).

Exploring the natural world together can help foster relationships as well as developing physical skills.

> *Children talk about the features of their own immediate environment and how environments might vary from one another.*
>
> Understanding the world, Development Matters, 2014

When taking part in science experiences, children have many interesting ways of connecting with and contributing to the natural world. An important subsection of the outcome for Understanding the world is 'Children become socially responsible and show respect for the environment'. For example, a toddler watering a plant is beginning to form knowledge about plants needing water.

> *Give opportunities to design practical, attractive environments, for example, taking care of the flowerbeds or organising equipment outdoors.*
>
> Understanding the world: The world, Enabling environments (40–60+ months) Development Matters, 2014

Children are connected with and contribute to their world when they:

- ✓ develop a respect for the natural environment
- ✓ investigate the interdependence of living things
- ✓ work together on a science project
- ✓ explore science concepts through play
- ✓ explore, infer, predict and hypothesise as they engage in science-related play
- ✓ learn about the interdependence of land, people, plants and animals during science experiences
- ✓ learn to care for the natural environment
- ✓ explore living and non-living things
- ✓ notice and respond to change
- ✓ explore the impact of humans on environments

Smelling freshly–picked parsley offers experience of biological science, using the senses of smell, sight and touch.

Developing a sense of well-being

Many science experiences offer children opportunities to develop personal and social skills and attitudes:

> *Provide areas to mirror different moods and feelings – quiet restful areas as well as areas for active exploration.*
>
> Personal, social and emotional development: Managing feelings and behaviour, Enabling environments (22–36 months) Development Matters, 2014

Children have a strong sense of well-being when they:

- ✓ communicate their bodily needs
- ✓ develop greater awareness of what their bodies can do (biological science)
- ✓ develop simple chemical science knowledge as they learn to use soap and water to wash their hands
- ✓ develop their fine motor skills as they record information about science
- ✓ develop an awareness of biological science knowledge as they learn about healthy habits such as enjoying eating fruit and vegetables they have grown in the garden
- ✓ manipulate science equipment and tools with control
- ✓ develop scientific understanding by exploring their world through their senses
- ✓ develop spatial awareness as they explore their world (physical science)
- ✓ learn about physical science as they experiment with gross motor movements such as moving down a slide (beginning to understand gravity)

Children are confident and involved learners

Practitioners who see children as strong and capable learners – ready and willing to explore and learn about their world – are able to provide support for children to develop their understanding of science-related areas of learning.

Practitioners can show interest and encourage babies to explore simple science activities by stimulating their senses and encouraging active motor exploration of materials. Smelling and tasting a leaf, poking it with tiny fingers, feeling its smoothness or furriness all help a baby to build the concept of what a leaf is.

As well as their five senses, older children use more advanced skills of language and reasoning, enabling them to develop a more detailed understanding of the natural world. Practitioners can offer open-ended, loose materials and also ask questions to extend thinking. When older children are encouraged to persist with challenges and explore their ideas, they are engaging with science in a meaningful way.

Children are confident and involved learners when they:

- ✓ express wonder about and interest in science

- ✓ develop curiosity and creativity when engaged in science activities

- ✓ transfer and adapt their understanding of science

- ✓ develop an understanding about themselves through biological science

- ✓ use hands-on learning during science experiences

- ✓ develop concentration and persistence when pursuing a science project

- ✓ learn to enquire when researching or exploring a science concept

- ✓ predict and generalise about scientific ideas

- ✓ think reflectively about a scientific experience

- ✓ explore tools and technology as they learn about science

Children are effective communicators

Children use verbal and non-verbal language to communicate their ideas and understanding about science.

Using a wooden stick to bang recycled materials leads this child to respond verbally to what she sees and hears.

Children are effective communicators when they:

- ✓ use gestures, sounds and language related to their scientific learning

- ✓ exchange ideas, thoughts, questions and feelings about science

- ✓ sing and chant songs or rhymes related to their scientific understanding

- ✓ use their home language to label scientific objects and concepts

- ✓ use electronic and print media when exploring scientific activities

- ✓ incorporate mathematical understanding in scientific problem-solving

Chapter 3

How to support a child's sense of wonder

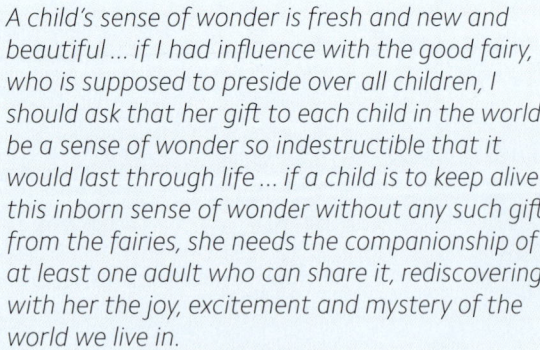

> A child's sense of wonder is fresh and new and beautiful ... if I had influence with the good fairy, who is supposed to preside over all children, I should ask that her gift to each child in the world be a sense of wonder so indestructible that it would last through life ... if a child is to keep alive this inborn sense of wonder without any such gift from the fairies, she needs the companionship of at least one adult who can share it, rediscovering with her the joy, excitement and mystery of the world we live in.
>
> Rachel Carson

When adults are in touch with the wonder of the natural world, they honour and support a child's sense of wonder about science.

Intentional teaching

Intentional teaching is one of the key aspects of practice in the early years. Practitioners who are 'intentional teachers' are deliberate, purposeful and thoughtful when they act or make decisions. This is not to be confused with rote teaching or doing things because they have always been done a certain way.

In her book *The Intentional Teacher*, Ann Epstein summarises this as:

> *Child-guided experience + adult-guided experience = optimal learning*
>
> *An effective early childhood programme combines both child-guided and adult-guided educational experiences. The terms 'child-guided experience' and 'adult-guided experience' do not refer to extremes (that is, they are not highly child-controlled or adult-controlled). Rather, adults play intentional roles in child-guided experience; and children have significant, active roles in adult-guided experience. Each takes advantage of planned or spontaneous, unexpected learning opportunities.*

In the UK, such methods are described in these ways:

> *Practitioners teach children by ensuring challenging, playful opportunities across the prime and specific areas of learning and development… closely matching what they provide to a child's current needs.*
>
> Development Matters, 2014
>
> *There must be a balance between structured learning through child-initiated activities and those directed by practitioners.*
>
> Framework for the Early Years, Wales

A beautiful flower that has dropped to the ground might at first be smelt, then handled and treasured. The practitioner could then provide resources for the child to translate what she has learnt into an observational drawing of her precious find, and remain available to discuss the child's learning about the properties of her special flower.

Intentional teaching involves deliberately setting up learning environments to support specific science experiences, as well as thoughtfully recognising and taking advantage of spontaneous teaching opportunities when they happen. An intentional teacher sets up both teacher-guided and child-guided experiences.

Asking questions is part of intentional teaching. Intentional teachers use social settings to help engage children in meaningful experiences. They model curiosity and open-ended thinking, for example, 'I wonder why?' 'How might?' This is often referred to as 'sustained shared thinking', where adults and children work together to solve problems, explore experiences and think critically about these experiences.

Practitioners can incorporate intentional teaching through planned experiences or through incidental teaching when an opportunity arises.

How does intentional teaching support scientific learning?

Intentional teachers have specific science learning goals that are also open-ended and allow children to develop a line of enquiry. These practitioners provide rich science learning environments with opportunities for creative science inquiry. They encourage children to explore through open-ended materials and intentionally guide children's learning without taking over.

> *Every time we teach a child something we keep him from inventing it himself ... That which we allow him to discover himself will remain with him.*
>
> **Jean Piaget**

Intentional teachers believe that children are strong and capable learners. They encourage scientific thinking through open-ended questioning that inspires curiosity and wonder. Intentional teachers model scientific thinking and problem-solving without solving the problems for the children. They decide when to enter into children's play situations and when to take the lead; they also know when to sit back.

Intentional teaching ideas for scientific learning

In the chapters that follow, we have made many suggestions of how to set up intentional science teaching experiences that specifically address each of the areas of science. These ideas are based on actual learning environments, and we hope some of these ideas might spark your interest or lead to an adaptation for your own setting and unique group of children.

All practitioners can teach science

Intentional teachers learn more about some of the basic concepts of science and simple scientific terminology by researching relevant areas of science in their preparation for teaching and learning.

Practitioners can also learn alongside the children and look up information with them as part of the investigation process. At the end of this book you will find useful resources for both practitioners and children, and web addresses of relevant science sites and organisations where you can find support, ideas and information – see p.90.

It is important to provide inspiration for children. When we notice the beauty of the world around us, we can use this to feed children's curiosity and encourage their creativity. We need to use this enjoyment to enthuse the children by sharing our own experiences with them.

We need a sense of wonder, too!

When practitioners learn about the natural world, they can become interested in all the different aspects of science, for example about how objects move, about the mystery of gravity, or about the wonder of natural phenomena. It is exciting to learn how plants and animals grow and relate to the natural environment, and it is even more important these days to understand and model how to 'Rethink, Reduce, Re-use and Recycle' in order to preserve our environment for the future.

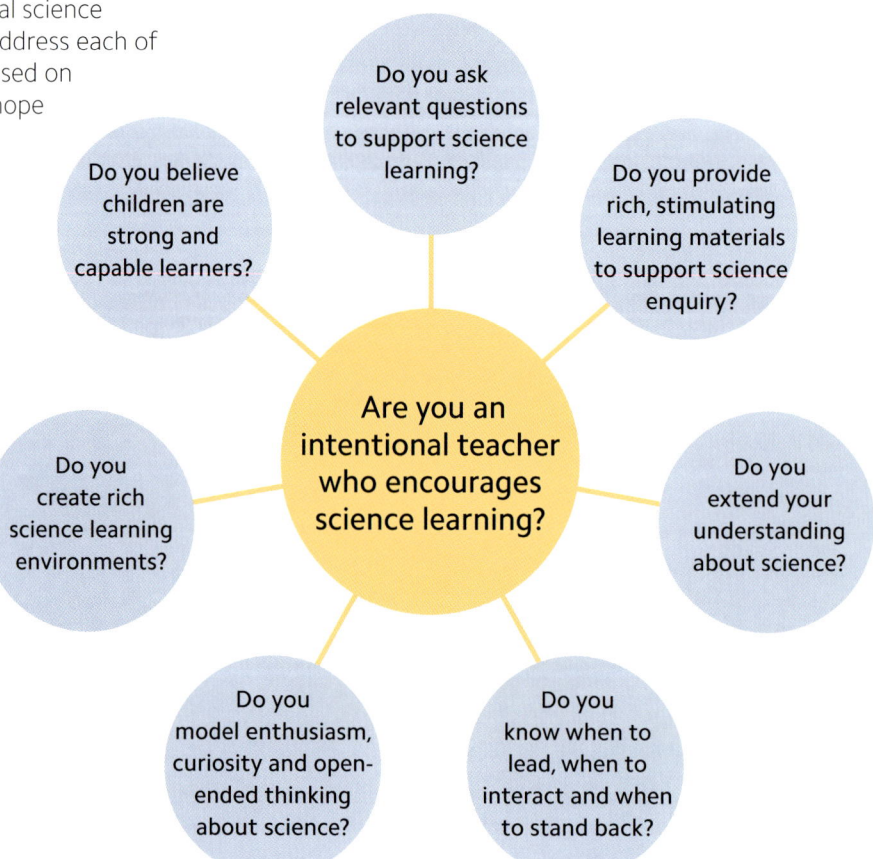

Developing dispositions for learning

Dispositions are characteristics that encourage each child to respond to learning opportunities. Nine key dispositions for learning have been identified, and these also relate to the learning of science. The dispositions are: curiosity, cooperation, confidence, creativity, commitment, enthusiasm, persistence, imagination and reflection.

1. Curiosity

Children are naturally curious. As adults we need to notice and not judge. We need to know when to stay quiet and when to speak. Staying quiet lets children focus on the experience and use all their senses to absorb information. We can then model using scientific vocabulary as we wonder aloud about what we are looking at. Children need the words to describe what they see and to show what they have understood about their experiences.

2. Cooperation

Many scientific enquiries require a group of children to work together towards a single outcome. Cooperation means working with others, listening to their ideas, rejoicing in their group achievements, sharing work and taking turns.

3. Confidence

Young children usually have confidence in their own abilities, thoughts and ideas. It is important to build on this disposition for learning by encouraging children to get involved in scientific experiences. Where children are less adventurous, practitioners should be available to support them, making gentle contributions and encouraging further involvement to extend their interest.

4. Creativity

Children are naturally creative, and expressing creativity reflects a child's feelings, ideas and imagination. Creativity involves exploring and expressing original ideas, and we should encourage children to pursue their ideas about science and work out answers by investigating possibilities.

Children can come up with exciting and unique answers to their own questions. For example, how could we save the rainwater to use in the vegetable patch? The answer could involve many creative solutions that need to be investigated to work out what might happen. How can we fix a bucket to collect the water running off a shed roof? Will sticky tape hold the downpipe onto the bucket? Children need opportunities to trial their suggestions and work out new solutions if necessary.

This baby grasps, looks at, mouths and smells the lemon as he investigates an unfamiliar object that he found on the mat. Freedom to explore supports his disposition of curiosity.

Encouraging the disposition of cooperation: by working together, the children and practitioners have constructed a scarecrow from recycled materials to put in the children's vegetable garden.

5. Commitment

Practitioners will encourage children's commitment to learning about science if they allow extended time to explore an idea without interruption. The provision of rich, stimulating and aesthetically appealing science-based learning opportunities that take time to achieve will entice children to commit their time and energy. Practitioners need to support children's interests with open-ended questioning, offering additional materials to enable the learning to continue.

6. Enthusiasm

It is not hard to encourage the natural enthusiasm of young children. When we show interest and provide stimuli for children's scientific enquiry, then enthusiasm will follow. Challenging scientific enquiries lead to rich learning about scientific concepts and vocabulary.

7. Persistence

Through enquiry-based scientific experiences, children can develop persistence. Trying different solutions to problems such as 'How can I make the water move from one place to another?' can encourage children to persist in order to achieve a positive outcome. The satisfaction of achievement is wonderful to see in children who have persevered to complete a long-term science goal. An example would be the satisfaction and delight children derive from eating fruit and vegetables grown in their own vegetable plot.

Creative use of real resources can stimulate children's imagination and extend scientific learning.

8. Imagination

Young children have wonderful imaginations and can come up with exciting solutions to questions based on their current understanding of the world. It is also important to remember that children who are in Piaget's 'preoperational cognitive stage' might believe in ideas such as 'rocks have feelings' or that 'a person up in the clouds is blowing the leaves in the trees'. Sometimes we need to let children hold on to their magical beliefs about the wonders in the natural environment. When they are ready, they will ask the questions that lead to a more realistic understanding about the wonders of the natural world.

9. Reflection

Children demonstrate the ability to reflect when they look back on their learning and relationships. Through discussions about what has happened, practitioners can help children to review and consolidate the scientific concepts and vocabulary they are learning. One way of encouraging this disposition is to support children in documenting their learning in observational drawings and digital photographs, and providing opportunities to discuss these documents with adults and other children.

Enquiry-based learning

What is scientific enquiry?

Scientific enquiry is the process of exploration, research, investigation, problem-solving, examination, analysis, inspection, study, and asking questions.

Science itself is enquiry; it is discovery through a systematic process of exploring the properties of objects through hands-on experiences. Through enquiry, children are learning how to learn, developing the skills and abilities to last throughout a lifetime. Rather than the retention of facts and knowledge, an enquiry-based approach allows children to be at the centre of the activity and be personally connected to their own learning.

Practitioners who show surprise, interest, pleasure and delight in the simple happenings of everyday life will have a profound impact on a child's future.

ENVIRONMENTAL SCIENCE	PHYSICAL SCIENCE
How can we look after it?	What is hot? What moves?

SCIENTIFIC ENQUIRY
Where? What?
How? Why?

BIOLOGICAL SCIENCE	EARTH SCIENCE
What is alive? What grows?	What can we see? What does it feel like?

Enquiry can be initiated by a child or a practitioner:

CHILD
notices, wonders, shares

PRACTITIONER
engages interest, sets up experiences, poses questions

A wonder or interest emerges – e.g. 'I wonder how the red flower got in the middle of the white ones.'

The adult recognises scientific possibilities and poses questions

Child notices, wonders and shares

The adult engages the child's interest, sets up experiences and poses questions: 'What can we see...?' 'What do we want to find...?' 'What do we think?' 'What do we already know about this?' 'How can we explore it?'

The adult uses intentional teaching, considers links to the relevant Framework, encourages dispositions for learning and facilitates scientific process skills.

The adult reflects on the scientific learning.

Planning for scientific enquiry

When planning for scientific enquiry it is important to consider the community and environment in which the children live. It is crucial to follow the children's interests and guide their innate curiosity about the world around them.

However, many popular topics that are explored in early years settings do not directly relate to young children's everyday experience. Some of these are:

- Space travel

- Planets and solar systems

- Dinosaurs

- Rainforests

- Antarctic and Arctic animals and their environments

Practitioners may wish to reflect on the place of these topics in a science curriculum.

Science educator Karen Worth (2010) suggests that children's interests in these topics can be explored through dramatic or imaginative play and discussions using books and other visual media. She advises taking care not to allow these topics to replace young children's direct enquiry-based learning into their own immediate environment.

Early years practitioners know the children in their group, and so are in a position to make informed decisions about what is appropriate for their setting.

Some popular topics in science may not relate to young children's everyday experience; however, practitioners might choose to incorporate them into imaginative play.

Four-year-old Ahmed and his mother excitedly told Jo (Ahmed's practitioner) about the turtles they saw on a family holiday. In response, Jo set up an underwater area to extend on Ahmed's interest. Jo then helped Ahmed to add to his scientific vocabulary and understanding about turtle habitats through an exploration of the resources together.

Consider following interests that children show in their immediate environment. These might include:

- Watching and drawing snails that live in the garden

- Investigating plastic ducks moving down a water slide

- Investigating how their bodies make shadows

- Feeling the textures of natural objects

- Investigating how food they have grown in the vegetable garden changes as it cooks

Connecting with the child's world

Using children's interests and information about their community and family backgrounds helps to make learning more meaningful. Inviting family members to participate in science projects and taking children into relevant community settings can provide exciting starting points and extensions to science activities. Family experiences or special events can be used as ideas to enhance the science curriculum.

Family members could buy or just take home the produce children have grown in their vegetable garden. In this and other ways, families can be involved in their children's learning. Practitioners could also ask parents for ideas and resources to support their children's interests.

The importance of a play-based approach to science

Intentional teachers know that play nourishes children's development and learning. Play needs time, space and an inviting environment.

Through play, children create ideas about their world, solve problems, communicate and negotiate with others. The foundation of many scientific concepts is laid down as children wonder about what they are seeing and are enticed to explore.

Repeated actions in new situations in play help children to consolidate their knowledge, practise new skills and understand scientific concepts. When the play is meaningful to the learner they are more likely to be involved and to learn from it.

Problem-based play allows children to engage in more complex thinking to develop deeper understandings. When children direct their own play, they develop feelings of confidence and competence and develop mastery over their learning and ability.

Catering for scientists under three

Babies actively explore their environments through their senses. They use their mouths, skin, eyes, ears and motor skills to collect new information about the world. They start to make connections between these experiences as they interact with the environment, and through these experiences, lay the foundations for future scientific understanding.

Simple ideas for science experiences for under-threes can be found in chapters five to eight of this book.

Toddlers and young two-year-olds are very active scientists, increasingly mobile, very curious and wanting to become more independent. They want to find out how the world works. They will test and experiment endlessly with the properties of objects. Sensory play experiences and opportunities to get messy will assist their learning.

The early years practitioner plays a very important role in modelling appropriate language and vocabulary, labelling items and objects, setting up simple science experiences and supporting under threes to make sense of their scientific findings about their immediate world.

Modelling a sense of wonder with children over three

A practitioner who begins with an attitude of excited, shared discovery, values the enquiry process and is concerned about the development, expression and recording of children's ideas. When a practitioner criticises and corrects, tells rather than asks, slavishly follows a prescribed curriculum, or chooses the learning activities themselves, the result can be a lost opportunity to inspire the children and engage them in rich and meaningful learning about the wonderful and fascinating world around them.

This baby is exploring the texture of the rug. Through this experience, she is laying the foundation upon which to construct her physical science knowledge and understanding about the characteristics and properties of materials.

This child is allowed extended time to explore and try out new ideas in his play.

Approaches to learning

Over centuries, many theories of learning and development have emerged which inform present day ideas about teaching and learning.

Friedrich Froebel

As far back as the early 18th century, Friedrich Froebel, known as the 'father of kindergarten', likened the role of teacher to a gardener. In this garden the children unfolded like flowers with the teacher as a planter of seeds. Froebel believed that development occurred through self-activity and play.

This child is learning through active involvement by handling some vegetables. She is smelling them, touching them, and exploring their properties, perhaps preparing them for eating or cooking.

John Dewey

At the beginning of the 19th century, John Dewey developed the idea of child-centred curriculum. This was built around the interests of children, with the teacher using their knowledge of the child to plan stimulating problems to solve and promote active involvement in activities based around real-life situations.

Jean Piaget

The idea of exploration and active involvement is also central to Jean Piaget's theory of cognitive development: that learning is enhanced through hands-on experiences with the physical world. Through activity and collaboration with others, children continuously organise, structure and restructure their experience — constructing their own knowledge. This is described as constructivist theory. Piaget's work on developmental stages reminds practitioners of the importance of each child's age and stage of development when considering their capability to understand concepts.

These children are actively constructing their own learning about the nature of light and shadow by experimenting with the overhead projector. Their collaboration is an example of Piaget's constructivist theory.

Lev Vygotsky

Lev Vygotsky was an influence at the same time as Piaget; he recognised the importance of language and social development in children's learning. Vygotsky developed socio-cultural theory, which has a focus on social interaction. Vygotsky stated that 'Learning awakens a variety of developmental processes that are able to operate only when the child is interacting with people in his environment and in collaboration with his peers.'

Abraham Maslow and Erik Erikson

Later, Abraham Maslow and Erik Erikson recognised the importance of children's emotional needs and feelings of self-worth and self-esteem in the learning process. They acknowledged the importance of a practitioner's ability to provide feedback, support and acceptance of children's efforts.

Present day programmes

Early years programmes have continued to develop around child-centred, socio-cultural educational ideas and theories developed over the years. Reggio Emilia, Te Whāriki (New Zealand), Steiner and Montessori approaches continue to provide inspiration for exemplary practice, and a foundation for recent curriculum frameworks in the UK and across the world.

In recent times, the importance of experiences out of doors, and of caring for and sustaining our environment, have resulted in the development of innovative early years programmes such as International Forest Schools and Nature Kindergartens.

Putting theory into practice

As children gather sensory information from their surroundings, they begin to construct their own ideas and make sense of the world. Often they will have misconceptions or only partial understanding of it. Many concepts are too complex for young children to initially gain a complete understanding. However, although they may not develop complex scientific ideas, their exploration can provide a basis for later concept development.

We can help children clarify their understanding by providing relevant experiences and asking carefully chosen open-ended questions to focus attention and encourage reflection. Young children can explore and discuss basic scientific ideas and begin to use scientific language. With support, they can direct their own enquiries and experiments, and begin to understand their world.

When children encounter exciting new things, practitioners can encourage them to take photos, draw and model what they see and record their conversations, resulting in new learning.

Programmes set in the outdoors provide children with exciting challenges and an opportunity to learn about the natural environment first hand using all of their senses.

Documenting children's learning

Recording a child's scientific understanding is a very important part of the curriculum planning cycle. Ways of making learning visible can include taking photographs, collecting observational drawings, learning stories, making models or implementing written observations. Interpretation of this information can help us to gain an insight into what each child is learning about science.

Scientific learning is a complex process, and it is helpful to document a child's learning across the year rather than making a single assessment, which would limit the learning to one moment rather than over an extended period of time. Regularly collecting evidence helps the practitioner to record the development of scientific learning.

Making learning visible

Photographs: selected photographs that clearly capture children as they delight in their learning can provide valuable records. Children can participate by taking some of the photographs and in choosing what to photograph and what to keep. Children's comments make a valuable addition as they explain what they were thinking or doing as the photo was taken.

Children's conversations: a record of a child's conversation can provide in-depth evidence, particularly when open-ended questioning invites the child to share ideas and understandings.

Displays and documentation activities: work samples, recorded science-related conversations with the children, photographs of the children during the scientific enquiry and any charts or anecdotal evidence that the practitioner has written can all contribute to the knowledge of each child's learning. Display the evidence on a board. You could use an existing notice-board, make a special panel or hang a series of signs from the ceiling. Share information with the children by ensuring that documentation is placed at the children's level and invite children to share their findings with their families. Questions that focus on scientific learning can be added to your displays to encourage family members to talk about what they see.

Attractively displayed documentation panels make the learning visible, and help to involve parents and families.

Learning stories: these are records of children as they engage in either individual or social learning contexts. Learning stories are a tool developed by Margaret Carr for use in the new Zealand Early Years Framework — Te Whāriki. A commonly used format for the learning story is the notice–reflect– respond format, which is as follows.

Drawings: a child's drawing can be a useful source of information about their scientific understanding. This can be more clearly recorded when, with the child's permission, the child's dictated comments are added to their drawing. Observational drawings are particularly useful and interesting, as they enable children to document what they see in a scientific experience, or experiment.

A learning story

Notice — **The story**

Describe in detail what the child says and does during the scientific experience.

Photograph(s)

Include one or more photograph that shows clearly what is happening and what the learning is.

Reflect — What does the learning story tell you about the child?

Identify and reflect on any significant scientific learning or interest shown by the child.

Link the learning to one or more outcome, customising the wording so it clearly describes the child's specific learning.

Link to the relevant learning goals.

Respond — **Learning opportunities**

What do you plan to do now? Explain how you will extend the child's scientific understanding/development/interests through intentional teaching, planned experiences or support from adults or additional resources, spaces, time or other organisation. Include the child's ideas where appropriate.

Describe in detail what the child says or does in response.

Chapter 4

Creating inspiring spaces for science

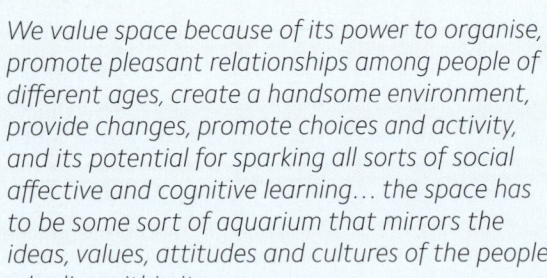

> *We value space because of its power to organise, promote pleasant relationships among people of different ages, create a handsome environment, provide changes, promote choices and activity, and its potential for sparking all sorts of social affective and cognitive learning... the space has to be some sort of aquarium that mirrors the ideas, values, attitudes and cultures of the people who live within it.*
>
> Loris Malaguzzi

Our changing world

Every child comes from a unique family, and in early years settings we have a rich variety of cultural identities that need to be acknowledged and embraced. For some children, time is shared between the two homes of separated parents, and any activities done in free time are often fast paced and sometimes tightly scheduled.

Early years settings can provide spaces where each child can take time to explore, think, experiment and learn. Children will flourish with the undivided attention of early years practitioners who know how to support their individual needs in an unhurried, interesting and nurturing environment. In the Reggio Emilia early years programmes in Italy, the practitioners value close connections with children and their families when planning and providing carefully designed early years programmes.

> *The structures, choices of materials and attractive ways in which educators set them up for the children become an open invitation to explore… the teachers pick up the children's interests and their ideas, share and discuss them among colleagues and then return them to the children themselves, engaging them in dialogue and offering tools, materials and strategies connected with the organisation of space to extend those ideas, to combine them or transform them.*
>
> **Edwards et al, 1998**

Children can explore physical science ideas through this imaginative play experience, as well as becoming familiar with the colours and textures of the natural world.

Working out how to pump water into the dry stream bed is an important physical science investigation for children.

Science across the curriculum

Scientific learning is perhaps at its best when children begin by expressing their wonder and interest in their world.

The discovery of a spider web in the playground might first involve shared language about what it is. The speculation about where it came from could lead to discussion about children's prior experiences with spiders, and what else they can find out. Early years practitioners will recognise the curriculum connections which can be made from this one discovery, not only in science but in areas such as language, literacy, music, movement, visual arts, maths and drama.

Children can be invited to look closely at the spider web and then to represent what they see through writing, drawing, modelling and art. The children, with their teachers, can then make decisions about how the learning progresses from this initial discovery:

- 'What do we know about this spider web?'

- 'What else do we want to know about the web?'

- 'How can we find out more?'

After the investigation, children can be asked:

- 'What have we learned about spider webs?'

The practitioner can do simple research to enable the learning to progress, and to retain the spark of wonder – building on the initial discovery, preparing resources, considering the possible science vocabulary and incorporating books, music, drama, art and maths.

> ## Organisation of materials in Reggio Emilia programmes
>
> *The environment is seen here as educating the child; in fact it is considered as 'the third educator' along with the team of teachers. In order to act as an educator for the child, the environment has to be flexible, that means that all the things that surround the children - the objects, the materials, and the structures are seen not as passive elements but on the contrary as elements that condition and are conditioned by the actions of children and adults who are active in it.*
>
> **Edwards et al., 1998**

Resources to support scientific enquiry

On the following pages we have included a range of science resources to support children's scientific learning. Many basic items are relevant to other parts of the curriculum, such as large pieces of fabric, interesting containers in which to display items and resources for recording children's learning.

Visual arts project idea – a spider web made from sticks and wool.

Systematic storage of resources in transparent containers means that children and practitioners can locate materials easily and contributes to the aesthetics of the environment.

Scientific resources

Fabric: large pieces of fabric can define areas and can be draped to cover surfaces or objects. Aim for natural colours – blacks, blues, greens, earth colours. Include thick, softly surfaced fabrics, differently textured fabrics and light, filmy fabrics.

Observational and measuring equipment: good quality magnifying glasses, mirrors, light boxes, rain gauges, rulers, tape measures, scales, clocks, timers, thermometers, droppers, pipettes and plastic tubing are all very useful resources to have in your setting.

Natural materials: stones, rocks, gravel, mud, sand, earth, water, sticks, driftwood, logs, pinecones, shells, cleaned bones, birds' nests, feathers, cleaned and blown eggshells, insect skeletons, seed pods, seeds, seedlings, edible plants such as herbs, vegetables and fruit, and easy-to-grow non-toxic plants such as geraniums.

Interesting containers: tubs, trays, aquariums, cages, buckets, paper cups, bags, baskets and bowls of various shapes and sizes made from substances other than plastic (e.g. wood, glass, metal, cane, raffia, wire, fabric, reeds or clay).

Fasteners: rope, string, tape (sticky, masking, duct), staplers, glues, fine wire, elastic fasteners, cable ties and clips and pegs of all sorts and sizes.

Recycled products: building and plumbing off-cuts, (e.g. plastic drainpipes and guttering), tubing, plastic drink and milk containers, funnels and holders, wood off-cuts, recycled plastic bottles and jars, egg boxes, cardboard boxes, cardboard cylinders, used office paper, empty spools, boxes, recycled or reject objects from manufacturers, carpet off-cuts, seed-growing containers and plant pots.

Mark making and writing resources: lead pencils, coloured pencils, highlighters, inks, paints, digital cameras, tablets and computers to record science findings.

Paper: white and coloured paper of varying thicknesses and sizes, different colours of cellophane, greaseproof paper, tracing paper, blotting paper, foil, cling film and paint shade cards.

Household and garage items: keys, padlocks, nuts, bolts, screws, plastic and metal fittings, handles, cogs, magnets, child-sized hammers, nails with large heads, sandpaper and sandpaper blocks, hand drills, safety goggles, ear defenders, gloves, pliers, clamps, small watering-cans, wellingtons, child-sized metal spades, trowels and garden forks.

Reference materials: science-related picture books, storybooks, simple factual books, picture and simple word dictionaries, gardening magazines, posters and the internet sites.

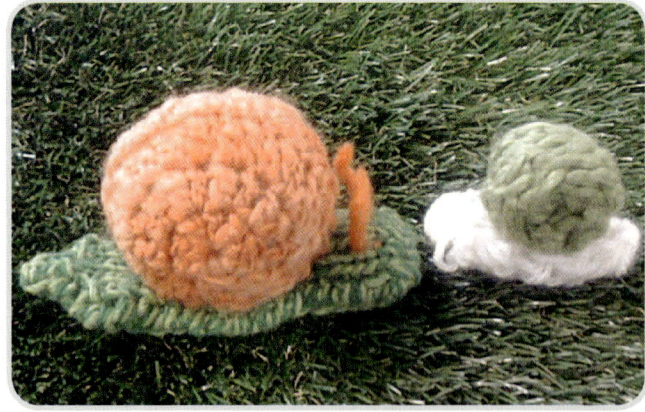

Small world props: animal and insect small world plastic toys, finger and hand puppets, dolls, small world people from varying cultural backgrounds.

Creating aesthetically pleasing environments

Environments affect our behaviour

The physical and social environment experienced in an early years setting has a powerful effect on the behaviour of children and practitioners. Early years educator Jim Greenman explores this concept in detail. He explains that spaces contain sensory information which stimulate the brain. The colour, smell, sound and complexity of the environment affect our emotional reactions. Our environment influences the hormones we produce, our blood pressure, our rate of breathing, the tension in our muscles and even our digestion. Greenman asserts that the quality and type of objects in the environment are important because the shape, texture and even the smell of an object directly affect us.

> A smooth stone, a soft velvet cloth, a bumpy pinecone, a rounded piece of wood, the soft, leather covering on a cushion – all of these tempt you to touch and smell them.

> A cold, hard concrete step, a piece of harsh synthetic material, a cracked plastic cup – these do not invite you to touch or handle them.

Creating an invitation to science learning

When presenting scientific experiences to children, consideration should be given to the open-endedness of the materials chosen and the settings into which the materials are placed. Consider how this environment will invite every child to explore and enquire; how it will provoke them to wonder, to concentrate and persevere.

Loose materials that can be moved and used in an open-ended manner invite children to make decisions about how to explore the scientific possibilities in this outdoor space.

Children learn through their senses

Children need to use all five senses when exploring and learning. When we solely rely on television, tablets, DVDs and computer games as experiences for children we severely limit sensory input and learning.

Looking at a real tree, smelling it, listening to its branches as they sway in the wind, feeling its rough bark and maybe tasting a leaf will enable a child to build a more complete picture of a tree than looking at it in a book or on a computer screen.

Using real resources

Where possible, investigate real situations using real tools. When real resources are used in a thoughtfully constructed environment the science learning can be richer. Children's investigations will be more independent and their attention spans are likely to be longer if the tools and materials available are appropriate. A plastic spade will be a frustrating tool to use to dig in real earth. A metal child-sized spade will encourage the child to persist because it will cut through the soil more easily. Magnifying glasses that actually increase the size of the viewed item so fine detail is visible are bound to encourage children to observe for longer and to explore further than a cheaper, less effective tool.

A collection of inspiring natural materials to use in science experiences provides children with relevant creative and authentic learning opportunities. For example, sorting shells and exploring their differences appeals to the senses. The smoothness of a shell and its natural shape and colours are more interesting to explore than commercially available plastic sorting objects.

Reflecting the child's views and voice

When setting up and supporting science activities it is important to create environments that reflect the child's views and allow for each child's voice to be heard. Use information learnt from listening to children and from the documentation of their interests and current knowledge of scientific as the basis of further science enquiries.

These children are exploring the scientific concepts of floating and sinking using props that relate to their current interest in alligators.

Safety considerations

Health and safety is an important consideration when planning for science activities. Practitioners must see safety as part of the learning process, and while allowing children the opportunity to fully explore the natural world they must strike a sensible balance between free experimentation and teaching children the safety considerations. Washing hands after handling natural materials and safely handling science equipment, tools, mechanical objects, and sharp and heavy objects are all part of learning about science. Putting on gloves, goggles or ear protectors can provide an experience in itself, especially when children understand the reasons why we need to wear these items.

For additional guidance, see page 96 for a list of resources and contacts to help with Health and Safety in science.

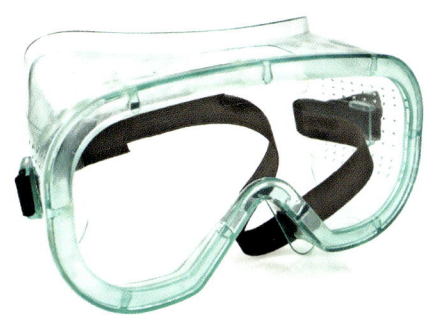

Some possible science learning safety considerations

Birds	Some children have allergies to birds, so handle birds' nests carefully – they can contain bird mites; moist seeds and bird droppings can cause diseases, so keep bird cages and food containers clean, and replace the food regularly.
Insects	Some children are allergic to insect stings – be aware of insect nests, and keep an up-to-date list of vulnerable children and their medication.
Plants	Some plants are toxic, and some children have allergies to certain plants, so this should be considered before touching or tasting is encouraged; ensure leaves do not irritate the skin; teach children to be wary of any insects found on plants.
Rocks	These contain many minerals and should not be put in the mouth.
Soil	Soil can contain contaminants such as cat faeces, poisons and air-borne contaminants – it is important to ensure potting mix is kept moist so children do not inhale the dust it produces. Buy sterilised soil for playing and planting.
Sun	We need to protect children from harmful rays of the sun on the skin and on the retina of the eyes. Monitor their exposure to the sun, especially in the middle of the day, and ensure they wear hats and sunscreen on visits and trips. Do not look directly at the sun; remind children not to look at the sun if it peeks out from behind a cloud.
Tools and mechanical items	Choose items that are child-sized and safe for children to use, and then supervise closely. Resources such as pulleys, torches, metal spades, hammers and nails, saws, tweezers should all be carefully supervised, and only introduced after the children have been shown how to use them safely.
Water	Ensure water comes from a safe water supply and is not too hot or too cold. Encourage children to only drink or taste water from a clean water tap, water bottle or other safe source. Be careful when using ice, as this is not always made from clean water.
Wind	Wear goggles if exploring the effects of wind to avoid dust getting into the eyes. Make sure kite strings do not get tightly wound around fingers or wrists. Keep a safe distance from fans or rotating surfaces.

Chapter 5

Biological science

Biological science in the early years

What is biological science?

Biological science is the study of living things. In early years education we focus on the characteristics of plants and animals, including humans. We promote awareness that all living things within the environment need to be treated with care and respect.

Why is biological science important?

Young children are natural scientists. They are keen to explore, investigate and learn about living things within their world. Early years practitioners can help young children to develop an understanding of biological concepts through experiences with living things, such as gardening, and caring for pets and wildlife.

These valuable and rich experiences can help young children to respect and care for plants and animals. It also helps children to develop a sense of belonging, wellbeing and connection with their world.

Plants

Every time young children get involved in gardening, they learn how to nurture, appreciate and care for other living things.

Most young children enjoy helping in the garden, and love to dig and plant. The physical and sensory experience of working in the garden provides wonderful scientific learning opportunities.

The practitioner's role

Practitioners should:

- Demonstrate a personal interest in gardening and physical work, such as weeding, digging and planting.

- Share their enjoyment in caring for animals, such as pets and wildlife.

- Promote healthy well-being through healthy eating habits and physical activity in the outdoors, (e.g. by gardening, walking and working with the children to provide a rich environment for learning out of doors).

- Show that they are inspired by the beauty of nature, by sharing experiences with children.

Some starting points for investigation

- Plants need water and food to grow.

- Plants change in appearance.

- There are many different types of plants.

- Plants can have stems, leaves, flowers and roots.

- Seeds grow into plants.

Scientific learning possibilities

- Caring for plants — trees, vegetables, herbs and flowers.

- Caring for wildflowers and the endangered species.

- Developing an understanding about plant life cycles and food chains.

- Developing an appreciation for the beauty and value of the environment.

- Understanding the differences between living and non-living things.

- Finding differences and similarities between living things, e.g. plant identification.

Developing scientific process skills

Skills include observing, classifying, measuring, identifying, problem-solving, asking questions, predicating and using simple scientific terminology relating to plants and seeds.

Experiences for children from birth to 3 years

From a very young age, children are eager to explore their world using all their senses. Going for walks in the outdoor environment provides babies and toddlers with wonderful opportunities to experience nature.

★ Provide experiences that babies and toddlers can do themselves, such as growing their own seeds in small recyclable pots or yoghurt containers. Remember to grow seeds that are quick and easy to germinate e.g. mustard and cress seeds.

★ Get babies and toddlers to participate in gardening activities, such as planting and watering, picking, smelling and tasting.

★ Make gardening simple for children by choosing plants that are easy to grow.

★ Plant vegetables which are easy to harvest, e.g. cherry tomatoes, strawberries, beans, potatoes and sweet corn.

Experiences for children from 4 to 5 years

Most young children enjoy playing outdoors. Children need time to investigate their own natural environment. Experiences such as nature walks, treasure hunts and time just to enjoy being outside will help children to develop a deeper understanding of the living world.

★ Supply cameras, magnifying glasses, binoculars and bug collectors to help children to observe, explore and investigate in the outdoor environment.

★ Create simple experiences, such as a broad bean activity in which children can watch the seeds germinate and start to grow. They can also closely observe how roots form, which helps them to understand about the stages of a plant lifecycle.

★ Provide spaces indoors for children to observe, discuss, explore and investigate plant life.

★ Provide writing materials so children can document and record plant growth patterns.

★ Provide a wide variety of books including picture and factual books such as Christine Back's *Broad Bean*, which has wonderful photographic representation of the growth of a broad bean into a plant. Supply a variety of gardening magazines, seed packets, photographs and plant posters – these are all useful resources for young children to look at.

★ Organise regular market days where parents and staff can purchase the organic vegetables, fruit, herbs, bush food and flowers produced by the children.

★ Provide opportunities and materials for young children to express their thoughts, ideas and what they see through observational drawings.

★ Flowers, bulbs, seeds and plants can be used as inspiration for beautiful artwork. As young children explore their environment, encourage them to look closely at the shapes and textures of flowers and leaves.

★ After picking fresh food from the garden, children can enjoy helping to prepare and cook the harvest either at home or in your setting or school.

Having a garden in your setting is a wonderful learning resource.

Families are encouraged to share home grown produce at this Children's Centre.

Herb gardens

Growing herbs will encourage children to use and develop their senses of sight, smell, touch and taste. If space is a problem, herbs can be grown quite easily in pots or on a sunny windowsill.

- Plant herbs such as parsley, basil, thyme, oregano, mint, spearmint, lemon verbena and rosemary - not only are they easy to grow, but they can also be very fragrant.

- Herbs can be propagated (bred from a parent plant) – so go ahead and propagate! Rosemary is easy to grow – use short cuttings form the end of a branch.

- Share your herbs with parents, families, staff and wider community.

Flower gardens

A flower garden will give pleasure to children, adults and wildlife alike. It prompts imagination, visual stimulation and a sense of wonder. If you have small outdoor spaces you could use pots, wooden half barrels, recycled cans and tyres to plant flowers in.

Vegetable and fruit gardens

Growing your own vegetables and fruit will provide opportunities for young children to develop an understanding about where food comes from. Children also learn to care for and respect their environment by participating in garden activities such as digging, planting, watering and weeding.

- Plant fruit and vegetables such as pumpkins, strawberries, celery, beetroot, spinach and lettuces as they are easy to grow and pick.

Planting strawberries is a fun and fairly straightforward activity that will reap delicious, rewarding results!

Green areas

All children need green spaces; they need to feel real grass with their hands and under their feet. Even if you have a limited outdoor area, plan at least for a small grass patch.

- Mounds of grassed spaces provide opportunities for children to climb up and slide down. They also provide children with wonderful opportunities to develop their physical skills.

Outdoor play spaces provide opportunities for young children to climb, crawl and roll down sloping hills.

Indoor plants

Provide opportunities for young children to bring the outdoors inside by growing indoor plants. This allows children to take responsibility and care for their plants and it further enhances their learning about living things. Indoor plants also provide a sense of emotional and physical well-being. Real plants also enhance the aesthetics of the indoor learning environment.

Try growing onions on a window ledge in your setting.

At this Children's centre, various herbs and vegetables are grown in recycled tubs and tyres.

Safety note!

Check with your garden centre when you buy plants to avoid those that are poisonous.

Trees, including deciduous and fruit trees

Trees are magical. They provide mystery, enchantment, wonder, delight, shade, patterns, a sense of calmness and well-being – and life! Whatever the size of your outdoor environment, you can find a tree to suit your space.

- All children need time to explore, investigate, climb and observe trees. The simplest of outdoor experiences, such as having a picnic under a tree, going for a nature walk in your outdoor setting, or participating in an excursion to the local park can provide endless science learning opportunities.

- Organise excursions to local parks or gardens, and talk to children about the trees they can see, how many different types they can find and which ones they like the best.

- Read Shel Silverstein's *The Giving Tree* – a book about a boy and a tree that keeps giving. Discuss why trees are important to our environment.

- Plant deciduous trees so that children can observe changes that take place over the year.

- Provide opportunities for young children to practise their predicting and measuring skills by following the growth of a tree. They can predict how much the tree will grow and compare the height with other trees. Children can easily do this by making simple observations such as 'this tree is bigger/smaller than…'

This tree in a nursery garden provides many opportunities for exploration and imagination.

- Visit a local garden centre or organise a visit from a local gardener to your setting. Contact a specialist nursery to obtain further information about how to plant trees and what trees are best suited to your environment.

The children in this setting spend long periods of time observing, exploring and investigating their natural environment

Every early years setting will benefit from having at least one fruit tree, such as an apple or pear tree. Get young children involved in planting and caring for the tree, and harvesting the fruit.

> *Being outdoors has a positive impact on children's sense of well-being and helps all aspects of children's development... Outdoors environments offer children freedom to explore, use their senses, and be physically active and exuberant.*
>
> EYFS 2014

By planting trees with young children, you will create beauty, harmony and a sense of wellbeing. You will also create habitats for wildlife, as well as spaces for young children to explore, investigate, play, climb and appreciate the magic of living things in their environment.

> *He who plants a tree, plants a hope.*
>
> Lucy Larcom

Children need time to play, explore and connect with nature.

Animals

Many young children have an interest in and love of animals. Keeping pets and caring for wildlife within your early years setting can be a wonderful way for children to learn more about biological science. Children learn many life skills when caring for animals.

Some starting points for investigation

- Animals need water, food and air to survive.

- Animals have lifecycles.

- Animals need environments that meet their specific needs.

- There are many different types of animals.

Scientific learning possibilities

- The lifecycle of animals.

- Caring for animals.

- Differences between animals and their particular needs.

Developing an understanding of scientific processes

Skills include observing, asking questions and communicating their ideas and findings, predicting, problem-solving, comparing and classifying.

Other areas of learning

- Mathematical skills – measuring the amount of food and water required; counting the number of eggs collected.

- Sense of responsibility – caring for animals.

- Communication skills – speaking, discussing, listening and recording.

- Social skills – taking turns and sharing responsibilities.

- Environmental awareness.

Keeping chickens

Chickens are environmentally-friendly pets. Keeping and caring for chickens in early years settings can be exciting and fun. Pet chickens can also provide ways for young children to further develop their ability to care for a living creature. They can help to feed the chickens, collect eggs and assist with cleaning the hen house.

If your setting decides to keep chickens, involve the children and families in designing and building the hen house. This can provide many wonderful learning opportunities for all and will give the children a feeling of responsibility.

There are a few important things to consider before getting chickens at your setting:

- Check with your local council regarding regulations about keeping chickens.

- Housing construction needs to be cat and fox proof — security is vital!

- Chickens need to have a perch that gives them comfort and a place to sleep at night. Hens also need a nesting box for laying eggs.

- Different breeds: bantams are small, cheap to feed, child-friendly and good egg layers. Take advice on the right sort of hens for your setting.

For more information on keeping chickens refer to the links on page 96.

Safety note!

When caring for and keeping animals within an early years setting, it is important that animals are cared for properly and respectfully. Do your research!

Stick insects

Most young children are fascinated by bugs and insects. Stick insects are interesting pets to keep and are quite easy to care for. You can purchase stick insects from breeders or pet shops.

- If you decide to keep stick insects, you should get the children involved in planning, designing and building a home for them.

- Young children can help with feeding and caring for the stick insects. A stick insect diet consists simply of fresh leaves. Make sure that the leaves you feed your stick insects are free from any pesticides.

- Children can spray the leaves each day with cold water and replace the leaves once a week.

- Provide opportunities and a range of materials for young children to express their thoughts, ideas and observations through drawings, paintings and modelling with pipe cleaners and/or modelling clay.

For more information about caring for stick insects see page 96.

In this setting the parents helped create this stick insect home.

Snails

Searching for snails can be lots of fun! In cool and wet weather there are more opportunities for children to find snails in the outdoor environment. Encourage children to look under the rocks and plants in their snail hunt.

- Provide a range of posters, factual texts and picture books, so that young children are able to further investigate, study and learn.

Provide young children with various materials to allow them to closely observe and investigate snails.

Creative resources

It is not always possible to keep pets within your early years setting. However, you can still incorporate learning about animals by creating inviting play spaces, experiences and resources around a specific animal-based theme. For example, include puppets, soft toys, figurines, mark making materials, dressing-up clothes and face paints, posters and books that are easily accessible at a child level.

- Use natural and recycled materials to create wonderful play spaces, such as recycled timber, logs, twigs and fallen tree branches, leaves, cones, pods, plants in pots. Populate the spaces with small world animals or soft toys.

- Get children involved in making props and resources. Try this recipe to make your own 'eggs'. This recipe can easily be adapted – you could try making any size or type of egg, from bird's eggs to dinosaur eggs!

The egg recipe

You will need:

- 6 cups of plain flour

- 3 cups of used coffee granules

- Food colouring

- 3 cups of salt

- 3 cups of sand

- 4 cups of cold water

What to do:

Add all of the ingredients to a bowl and stir together.

Mix water into the dry ingredients until the mixture is like dough.

Roll a piece of dough the size of small avocado and mould it into an egg shape.

Bake for 20–30 minutes in an oven (120°C) or dry for 2 to 3 days in a warm indoor area and/or outdoors on a sunny day.

A nesting box in an early years garden.

Wildlife: birds

Create a wildlife friendly environment within your early years setting and involve young children in creating a habitat for many native animals and plants. Native plants will attract wildlife and birds.

- Provide safe places for bird wildlife by putting up nesting boxes in trees. Nesting boxes can be put up at any time of year. They should be high above the ground, protected from predators and weather conditions such as rain and severe sun.

- Provide birdbaths and hanging bird feeders so children can enjoy watching the native birds. This will provide many beautiful and exciting science learning opportunities.

- Do something that involves the wider community, such as sponsoring or adopting an endangered animal or species through your local zoo or a wildlife charity. See page 96 for organisations you may wish to look into.

In order to explore confidently, children need to be curious, open-minded, willing to concentrate and to feel that their ideas are worth investigating. While the world is familiar to you, your children are experiencing its wonder for the first time. Your enthusiasm for finding out about the world will be infectious – and will help them want to learn. Take a look at some of the books on things like liquids, materials, weather, the natural world and technology in the children's section of the library, or at some websites. They will help to remind you just how much science is around us all the time and give you an idea of the sort of questions that young children have about the world.

Linda Thornton and Pat Brunton, *Science All Around*, Early Education

Children know about similarities and differences in relation to places, objects, materials and living things. They talk about the features of their own immediate environment and how environments might vary from one another. They make observations of animals and plants and explain why some things occur, and talk about changes.

Understanding the world, Development Matters, 2014

The human body

Much of a young child's life is spent learning how to use and understand their bodies — from crawling, rolling over, walking, jumping, running, grasping objects, touching, tasting and smelling.

Developing an understanding about their bodies helps young children to understand their world.

● Help young children to develop a deeper understanding about the human body by providing safe and nurturing learning environments, rich experiences and resources, and give them time to explore, play and discover. Offer opportunities for young children to explore and use all their senses (sight, taste, smell, hearing and touch).

● Help young children to develop their listening skills by exploring different sounds in the environment. Discuss the sounds children can hear and encourage them to record their findings through drawing.

● Provide resources and materials for children to explore and use their senses, such as natural materials, plants, cameras, magnifying glasses, scales, paper, pencils, journals, scissors, measuring cups, glass jars, spoons, graters, mortar and pestle.

● Make perfume or perfumed soup with fragrant leaves, flowers and herbs, such as lemon verbena, mint and basil.

● Display freshly cut fragrant flowers in the classroom environment.

● Create scented playdough by using fresh herbs or essential oils.

● Design a sensory herb garden. If outdoor space is limited you could plant herbs and flowers in pots.

● Young children may wish to collect and dry the herbs or flowers that have a strong smell to make potpourri.

● There are many children's books about the senses that you could incorporate in your learning programme.

● Invite children to help with cooking activities. They can observe, taste, smell and feel the ingredients, prepare the meal and of course eat it. Provide recipe books and magazines with pictures of food. This also provides opportunities for children to understand where food comes from, the different types of food and the importance of eating a balanced diet.

● Prepare picnic hampers, include lots of different fruit and vegetables and enjoy eating in the great outdoors.

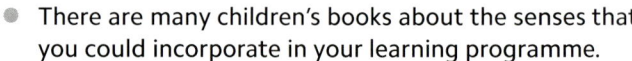

> *Children are confident to try new activities, and say why they like some activities more than others. They are confident to speak in a familiar group, will talk about their ideas, and will choose the resources they need for their chosen activities. They say when they do or don't need help.*
>
> Personal, social and emotional development, Self-confidence and self-awareness, EYFS, 2014

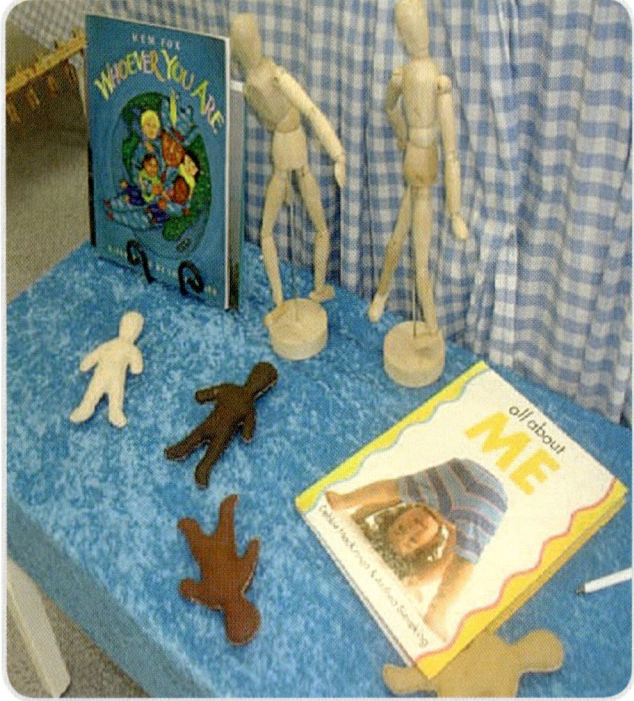

Providing resources such as books and props helps increase children's understanding of how their bodies work.

There are many other potential learning possibilities for young children to learn about themselves in every day experiences.

- Display photos of children, families and staff throughout the setting. This promotes a sense of belonging for the child, family and staff members.

- Place various mirrors around the setting, including outdoors. Babies and young children are drawn to mirrors and they enjoy exploring their own reflection. Mirrors provide opportunities for children to further develop an awareness of how their bodies work, who they are as individuals and their place in the world.

- Provide opportunities for 3- to 5-year-olds to explore their facial expressions by creating a quiet place for careful observation. Include mirrors, pictures of themselves and of different types of faces, as well as paper and pencils. Display children's self-portraits in picture frames or mounted on coloured paper.

- Allow plenty of opportunities for children to express their ideas through collage, clay, drawing and painting.

- Display posters, photographs, magazines and books that illustrate the human body.

- Keep a growth chart and encourage children to record their height. Read stories about measuring growth, such as *Titch* by Pat Hutchins.

- Integrate music, movement and dance within your programme. Sing action songs such as 'Heads, shoulders, knees and toes'. These fun experiences will allow young children to further develop an understanding of how their bodies work.

- Allow ample time throughout the day for physical activity, such as running, jumping, climbing, gardening and cleaning.

- Provide opportunities for children to express their feelings and thoughts by providing appropriate resources and imaginative play props.

- Add various props and resources to create themed role play areas, such as a hospital, where children can learn about their bodies.

- Include other resources within your learning environment to help children to develop an understanding of the human body, such as dolls, soft toys, puzzles, skeletons and bones (you can purchase bones from your local butcher but remember to wash, clean and sterilise with bleach).

- Young children need time to rest, think and reflect. Early years practitioners can do this by providing calm and cosy places for young children to enjoy time alone. Use Reggio Emilia inspired mirrors such as the Kaleidoscope mirror; add sheer and flowing fabric, books, pictures, small mirrors, soft cushions, blankets, soft toys, gentle music and lighting.

Chapter 6

Physical science

 Learn from yesterday, live for today, hope for tomorrow. The important thing is not to stop questioning.

Albert Einstein

Physical science in the early years

Physical science is the study of non-living materials and energy in the inorganic world. In early years, physical science does not need to be complicated or difficult and can be introduced through everyday experiences. Natural curiosity will inspire young children to discover and wonder about the world.

Children need time to examine the physical environment using their senses. As they become involved in relevant learning experiences they will gain firsthand knowledge of the physical world, which will support their concept development in the following areas:

- The nature of materials
- Physical and chemical changes
- Forces and movement of objects
- Energy

The practitioner's role

Practitioners need to have an understanding of basic scientific concepts to be able to recognise opportunities to promote children's understanding of these concepts. They can then extend interest and learning by:

- Providing a variety of safe, natural and manufactured materials for children to experience.
- Drawing children's attention to materials and encouraging them to investigate characteristics.
- Encouraging children to look for similarities and differences in materials.
- Using descriptive language when talking about materials, for example, words to describe shape, colour, texture.
- Asking open-ended questions to support creative thinking.
- Documenting children's observations and learning.
- Sharing the joy of cooking with young children and their families.

> *Play is a key opportunity for children to think creatively and flexibly, solve problems and link ideas. Establish the enabling conditions for rich play: space, time, flexible resources, choice, control, warm and supportive relationships.*
>
> Characteristics of Effective Learning, Creating and thinking critically, Development Matters, 2014

> *The whole of science is nothing more than a refinement of everyday thinking.*
>
> Albert Einstein

Some starting points for investigation

- Everyday materials are made up of a variety of substances.
- Materials exist in different states.
- Materials have particular characteristics.
- Materials can be classified in a variety of ways.
- Everyday materials can be changed in a variety of ways.
- Objects move in particular ways.
- Magnets attract some objects.

Scientific learning possibilities

Children will be able to identify links between their own world and relevant science learning by using their science process skills in everyday life.

- Finding differences between living and non-living objects.
- Finding differences between natural and man-made substances.
- Using materials in different ways.
- Identifying patterns in the environment.
- Noticing changes when substances are mixed together.
- Observing cause and effect relationships.
- Investigating how simple machines work.
- Exploring the characteristics of light, sound and magnets.

Developing science process skills

Skills include problem-solving, observing, predicting, investigating, measuring, researching and communicating.

The nature of materials

It is essential that children have access to a range of safe materials in the environment. Practitioners can provide experiences where children can explore materials in different states: solid objects such as wood, rocks, metal, plastic; liquids such as water; and gases in bubbles.

Focus children's attention on gathering information using their senses.

Natural resources are of particular value as they can provide a wide variety of sizes, shapes, weights, smells and textures for children to experience. (See chapter 7 for an exploration of earth materials.)

An assortment of materials with different characteristics can be set up in exploratory areas or integrated into other learning areas.

Open-ended experiences allow children to investigate and organise materials in their own way.

It is helpful to provide objects with opposite characteristics (for example, something rough and something smooth) as this will help to encourage children to identify differences.

Older children can compare, discuss and record similarities and differences in materials.

> *Use the language of thinking and learning: 'think', 'know', 'remember', 'forget', 'idea', 'makes sense', 'plan', 'learn', 'find out', 'confused', 'figure out', 'trying to do'.*
>
> Characteristics of Effective Learning,
> Creating and thinking critically,
> Development Matters, 2014

By providing a variety of open-ended resources for use in play, we can encourage children to use materials in innovative ways. They will be more likely to see different possibilities and solve problems if we support divergent thinking skills.

It is important for children to have opportunities to suggest, predict, experiment, see what works and what doesn't, discuss possibilities and try again. Open-ended questions can promote divergent thinking.

Experiences involving materials with different characteristics

- Make feely boxes using objects of different shapes, sizes and textures.

- Provide warm bean bags or pillows filled with different scents, such as lavender.

- At meal times, provide a range of food and discuss the tastes and smells.

- Include bubbles in children's play.

- Fill plastic bottles with interesting materials.

- Make pillows, floor mats and cards using a range of man-made materials with different textures.

- Allow children to sort objects that go together, for example matching socks.

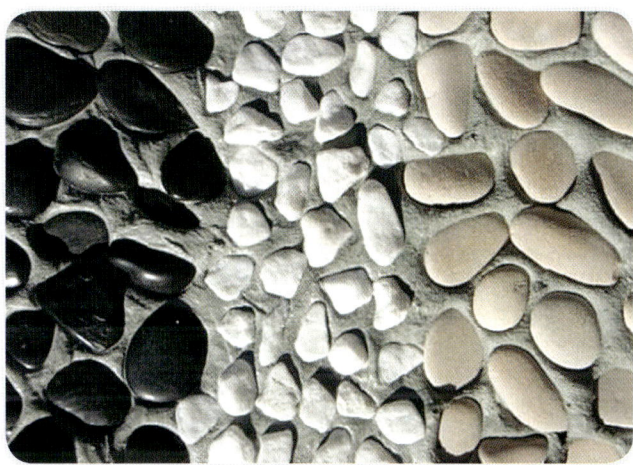

Stones set in mortar provide a safe way for young children to explore texture.

In this setting, practitioners have made use of handmade textured rugs to encourage young children to develop their sense of touch.

Include water play in different forms, for example water trays, water in the sand pit and digging area.

- Set up areas where children can investigate and record particular characteristics of materials, for example, objects that float/sink.

- Support young children to develop their problem-solving skills, for example, discuss how water can be stored.

- Provide opportunities for children to match objects in different ways, for example socks with shoes, knives with forks.

- Games provide opportunities for children to:
  - ★ use clues to identify the object
  - ★ identify the missing part of the object
  - ★ identify the smell
  - ★ identify the sound

- Provide a range of different scientific tools such as magnifying glasses to encourage children to develop their observation skills.

- Preschool children can focus their attention on particular qualities of materials and can sort and match objects based on differing attributes, for example: type of object (e.g. shell or stone), colours, size, texture, sound, weight, natural or man-made, waterproof or non-waterproof.

- Provide scales for children to investigate the weight of materials.

- Provide cameras for children to record their observations.

- Document observations and investigations in a variety of ways such as in books and on posters.

- Set up experiences in creative ways using recycled, open-ended props for play.

- Make dens using a variety of natural and recycled materials.

To encourage scientific thinking with older children, discuss:

- What can you do with water?

- What can you do with sand when it is wet and when it is dry?

- What can you do with recycled materials?

- What materials can you use to build a den?

- What materials can you use in your artwork?

> *In planning activities, ask yourself: Is this an opportunity for children to find their own ways to represent and develop their own ideas? Avoid children just reproducing someone else's ideas.*
>
> Characteristics of Effective Learning, Creating and thinking critically, Development Matters, 2014

Using recycled boxes to create dens and houses.

Physical change

Investigating changes in substances can be incorporated into children's play; it does not always need to take the form of an experiment. Practitioners can draw children's attention to the properties of materials and changes that we observe in substances whilst they play.

What is physical change?

In a physical change, the substance remains the same but may look different. Physical changes can occur when substances change shape or state, and these changes can often be reversed. They can be made by processes such as mixing, heating, melting, cooling, freezing and dissolving.

Experiences involving physical change

- Mix sand and water, soil and water, or sand and soil.

- Mix cornflour and water — 1 packet cornflour, 1 cup of water. Add food colouring.

- Mix soap and water — 1 cup of lux soap flakes and 1 cup of boiling water. Leave it to cool.

- Combine oil and coloured water — ¾ oil and ¼ coloured water in sealed bottles.

- Observe, discuss and record the different changes, for example when:
  - ★ water freezes
  - ★ ice warms in the sun
  - ★ clay is reconstituted

- Notice condensation forming on the windows.

- Inflate/deflate a basketball and observe and discuss the difference.

- Provide sponges in water play to explore absorption.

- Wash and dry clothes to explore evaporation.

- Paint outdoor paths with water and watch it dry in the sun.

> *Plan first-hand experiences and challenges appropriate to the development of the children.*
>
> Characteristics of Effective Learning, Playing and exploring, Engagement, Development Matters, 2014

Exploring different ways to make use of recycled water.

Using an assortment of materials for imaginative play.

- Set up areas in the sandpit and digging area with bowls, old saucepans and wooden spoons where children can mix materials.

- Paper-making involves a physical change. In this method there is no bleaching involved. Tear paper into pieces and place in a container. Cover with water and leave for about 24 hours. Combine the mixture together using a blender. Spread the mixture on to a wire mesh frame. (Old photo frames are useful and tulle can be used instead of wire.) Leave in the sun until dry. Once dry, remove from the frame.

Chemical change

Chemical science is concerned with understanding the composition and behaviour of substances. This area is not often mentioned when working with young children. However, chemistry is part of our everyday lives and includes the makeup of substances around us such as the colour of flowers, our own bodies, and items in our homes (for example, fridges, televisions, preservatives in food, batteries and medicines).

What is chemical change?

Chemical change occurs when a new substance is produced that has different physical and chemical properties. Matter is not destroyed or created in chemical change; however, the particles of the substance are rearranged. The substance produced during chemical change cannot be easily changed back into its original form.

Experiences involving chemical change

- Build and light fire in a safe outdoor environment. Through this experience children are able to:

  - collect and transport wood

  - work together as a team, sorting wood into lengths

  - explore maths concepts

  - keep warm and feel the success of getting the fire going

  - burn wood to make charcoal (this is an example of a chemical change).

Safety note!

Consideration needs to be given to how this can be achieved safely in your setting.

- Observe and record the changes in the colour of leaves.

- Observe the changes in metals left outdoors or place steel scouring pads in a little water and highlight the changes that occur.

- Observe the changes to rotting fruit and vegetables in the compost bin.

- Mix plaster of Paris and water (as per instructions on the packet) and use as a resource for play experiences.

> In planning activities, ask yourself: Is this an opportunity for children to find their own ways to represent and develop their own ideas? Avoid children just reproducing someone else's ideas.
>
> Characteristics of Effective Learning, Creating and thinking critically, Development Matters, 2014

- Drying fruit and vegetables in the sun includes both physical and chemical change.

Cooking experiences

- Cooking experiences are enjoyable and provide opportunities for children to discuss changes in substances. Chemical changes result in changes to flavour and texture.

- Cooking experiences that involve chemical changes include popping corn, and making pancakes, bread, scones, cheese and yoghurt.

Careful planning and implementation need to take place to make sure that cooking experiences are enjoyable and meaningful. Although children should not use ovens or stoves, they can be involved in the process by measuring, mixing, kneading, observing, discussing, predicting and documenting the process.

Making cheese

- 1 litre of full-cream milk

- 3 tablespoons lemon juice

- Salt and pepper

- Fresh herbs

Heat the milk in a saucepan until just before boiling point. Remove from the heat and add lemon juice. Cool and place the mixture into a very fine strainer or a piece of cheesecloth.

Allow to drain until all the whey has been removed from the mixture. Add salt and pepper to taste and any fresh herbs that you have grown in your garden.

Making cheese from yoghurt

- 1 litre of milk

- 3 tablespoons bioactive yoghurt (at room temperature)

Heat the milk in a saucepan until it froths and is just below boiling point (85°C). Cool until lukewarm (43°C).

Mix in the yoghurt. Cover and keep warm for 8–12 hours.

Refrigerate until the next day.

Add honey or fruits of your choice.

Once you have made yoghurt you can easily make cheese from it. Drain it in a very fine sieve or a clean cloth until the whey has completely dripped out. Add a little salt as needed, and fresh herbs or pepper.

Wholemeal bread

- 1 kg wholemeal flour

- 2 cups warm water

- 1 dessert spoon of honey

- 4 teaspoons dried yeast

- 2 tablespoons olive oil

- 2 teaspoons salt

Combine a little of the warm water and yeast, and then stir in the remaining water and honey. Allow to stand for about 10 minutes until the yeast bubbles.

Combine the flour and yeast mixture in a bowl and add the oil and salt.

Place onto a flat surface and knead for 10–15 minutes. The dough should become elastic in consistency.

Place the dough into a large oiled bowl and cover with plastic wrap and then a tea towel. Leave to rise for about 1 hour in a warm place.

Knead the dough again for about 5–10 minutes.

Shape into loaves or small bread rolls. Bake at 220°C for 20–40 minutes, depending on the size of the loaves.

Forces and movement of objects

Children of all ages can investigate the movement of objects in their environment. They can recognise that there is consistency in the movement of some objects, for example a toy will always drop to the ground if pushed over the side of the high chair. Children can experiment with objects moving in different directions and at different speeds.

Creating opportunities for children to experience cause and effect during their play will support further understanding in this area. Providing a range of equipment that can be pushed or pulled will allow children to learn about forces and movement.

> *Pay attention to how children engage in activities - the challenges faced, the effort, thought, learning and enjoyment. Talk more about the process than products.*
>
> Characteristics of Effective Learning, Playing and exploring, Development Matters, 2014

- Include equipment and toys that involve pushing, pulling and pedaling.

- Provide a variety of ways for children to observe and experiment with rolling and dropping objects (e.g. along the floor, in pipes, down inclines).

- Provide swings and slides.

Experiences involving force and movement

- Hang mobiles overhead for babies to watch and move with their hands or feet.

Large lightweight balls can be hung for children to push backwards and forward.

- Block play and making bridges is a way to explore balance and the movement of objects.

- Roll balls to children.

- Provide opportunities for children to move their bodies in different ways and talk about what they are doing.

- Set up a plank on an incline – which objects roll down the fastest?

- Provide balls of different sizes that are made from different materials – which ones bounce the highest?

- Now bounce balls on different surfaces – which surfaces make the balls bounce the highest?

- Set up interesting ways for children to explore moving objects.

- Explore the movement of water using lengths of guttering.

- Investigate the movement of objects in water and examine why some objects float and others sink.

- Make sail boats.

In this setting the practitioners help the children to make links between their scientific play and familiar stories.

- Include spinning tops in play.

- Investigate the tracks made by different objects and animals.

- Use funnels and waterwheels in water trays.

- Imaginative play set-ups can include ways to experience the movement of objects – for example, swings and pendulums.

- Investigate the work of the wind.

- Observe objects moving in the wind.

- Fly kites.

- Make parachutes and windmills.

Big sandpits out of doors provide wonderful opportunities for young children to explore movement.

Simple machines

Simple machines can provide further opportunities for children to investigate physical science. Simple machines can include wheels, pulleys, inclines, screws and levers. Children can explore and identify how machines can help to move objects.

- **Wheels:** What objects have wheels? How do wheels help objects move?

- **Slopes and inclines:** Provide slopes and inclines for children to investigate the movement of objects.

- **Screws, nuts and bolts:** Provide different sized screws, nuts and bolts for children to match and connect.

- **Levers:** Explore levers and include wheelbarrows to help children move objects.

- **Cogs:** Set up cogs and connective games for children to investigate.

- **Everyday objects:** Provide a range of these such as door knobs, keys and locks to allow children to investigate how things work.

- **Pumps:** Introduce these as part of water play.

- **Pulleys:** Simple pulleys can be safely set up indoors and outdoors to allow children to investigate how to move objects.

- **Woodwork:** This can provide opportunities for children to use tools.

Energy

Children's understanding of energy can develop through experiences with sound, light, heat and magnetism. Although children may not be able to fully grasp the concepts related to energy, they can identify the sun as a major source of heat and can experience different forms of energy.

Experiences involving energy

Heat

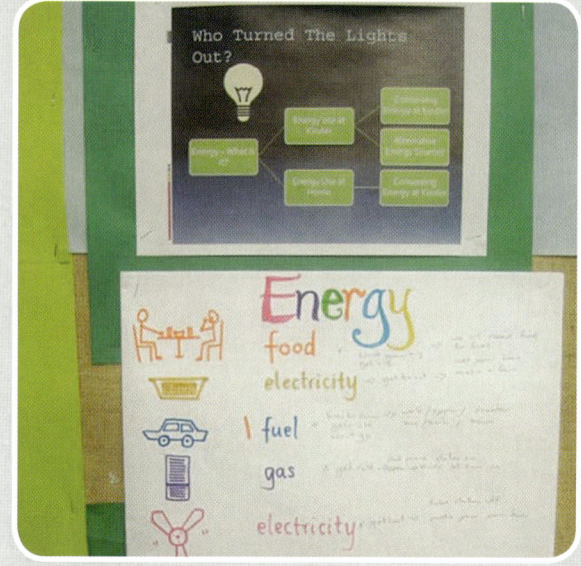

- Investigate and document sources of energy.

- Feel the warmth from the sun.

- Document objects that are hot and cold.

- Talk about how we feel the temperature through our skin, how we keep ourselves warm or cool, and the clothes that we wear in summer and winter.

- In colder weather, provide warm water in water play and in summer provide iced water.

- Provide warm and cold finger paint to make comparisons.

- Observe how heating and cooling causes changes to materials – for example, freezing water and melting butter.

- Melt ice in the sun and out of the sun – which melts faster?

- Include cooking experiences to observe changes when substances are heated.

Safety note!

Talk with children about the need to be careful with hot substances.

Light

- Include experiences with light in different forms – for example, sunlight, torches and electrical lights.

- Incorporate mirrors in children's play experiences.

- Talk about where we see reflections.

- Provide magnifying glasses for children to use.

- Introduce prisms for children to observe the separation of colours.

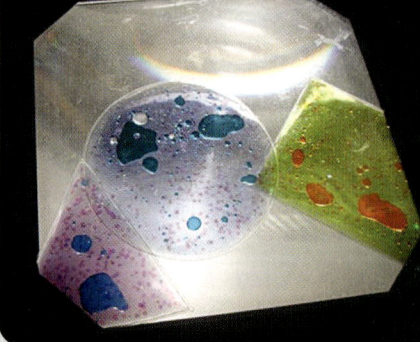

- Provide torches for use in dramatic play in darkened areas.

- Create puppet shadow boxes.

- Observe and record shadows made at different times of the day.

- Introduce binoculars for investigation.

- Make and use periscopes.

- Make use of overhead projectors.

- Use light boxes with a variety of transparent and non-transparent objects.

Sound

- Draw children's attention to objects in their environment that make sounds – for example, birds, cars, trucks, planes.

- Listen to and look at the wind blowing in the trees.

- Support children's understanding of how we hear with our ears.

- Make sounds, including talking and singing that are loud and soft – can children recognise the difference in volume?

- Encourage children to make soft and loud noises.

- Use tapes of sounds that are made by familiar objects in the environment – for example, kettles, washing machines, lawn mowers.

- Use a variety of recycled and natural materials to create musical instruments.

- Fill glass bottles with varying amounts of liquid to create a source for different pitched sounds.

- Provide a range of musical instruments for children to investigate sound.

- Feel the vibration made by a drum.

- Use a wide selection of recorded and live music for children to hear low and high pitch.

- Use tapes of animal sounds that children will recognise.

- Children can experiment with sound by using containers filled with different materials such as bells or rocks.

- Make wristbands with bells attached.

- Encourage children to make musical instruments using a variety of materials.

- Play games to explore sound such as:

  ★ Identify the voice

  ★ Identify the source of the sound

  ★ Identify the direction of the sound

Magnets

- Provide opportunities for children to experiment with magnets, and predict objects that will be attracted by them.

- Talk about how we can use magnets.

- Include cars, trucks and trains with magnets attached in play-based experiences.

- Provide puppets that can be moved with magnets.

- Make treasure chests filled with both magnetic and non-magnetic objects for children to sort using magnets.

Chapter 7

Earth science

 Those who contemplate the beauty of the earth, find reserves of strength that will endure as long as life lasts.

Rachel Carson

A SENSE OF WONDER

Earth science in the early years

Earth science is the study of the earth and its materials. In early years education it focuses on exploring and investigating the properties of non-living earth materials, such as water, soil, rocks, sand and mud. It also includes the sun and moon, and the influence weather plays on daily life.

It is important for young children to explore properties of earth materials through play-based experiences, rather than just learning facts about earth science.

The practitioner's role

Practitioners should:

- Provide an environment with appropriate materials, such as soil, sand, rocks and water, in which young children can investigate and explore earth science.

- Display interesting materials from the natural environment at child level.

- Demonstrate to children a respect and sense of wonder about the natural environment.

- Provide opportunities for children to document and record natural changes they have observed in their environment, such as taking photos, journal writing, drawing and three-dimensional representations.

Some stating points for investigation

- There are different forms of earth materials, for example, rocks, soils and water.

- Earth materials all have different characteristics.

- There are different weather conditions throughout the year which change our environment.

- The sun, moon and stars appear in the sky.

Scientific learning possibilities

- Observing and identifying different characteristics of rocks, sand, soil and water.

- Classifying and sorting a variety of different earth materials.

- Comparing, predicting and observing seasonal and weather changes.

- Predicting or comparing different clothes we wear in different weather conditions.

- Recording and discussing observations in relation to the movement of shadows, sun and moon.

Making simple observations about day and night.

Developing science process skills

Exploration will allow children to further develop their skills in comparing, classifying, predicting and communicating. Asking questions will encourage children to use simple scientific terminology about non-living things.

Benefits of water play

- Children can connect with nature.

- It provides a sensory play experience.

- It is soothing, calming, relaxing and refreshing.

Water

Water is vital for the survival of all living things. Most young children are intrigued by water and enjoy playing with it.

Water is a valuable and versatile resource and, as early years practitioners, you will need to provide learning opportunities for young children to develop an understanding of the importance of water, and how to respect and preserve it.

Experiences with water for children

- Many learning discoveries occur when young children are given opportunities to explore and play with water by simply splashing in puddles.

- Encourage young children to further investigate the properties of water by providing materials such as water troughs, small buckets, measuring cups, funnels and sponges.

In this setting, water is conserved by using recycled rainwater in play experiences and for watering the garden.

Rocks

The earth's surface is made up of different types of rocks. Rocks differ in their texture, colour, structure, size and shape. Collecting rocks provides opportunities for young children to explore their properties.

Experiences with rocks

- Create safe areas for babies and toddlers to observe and feel collections of smooth rocks.

- Create learning spaces that allow young children to further develop scientific process skills, such as sorting and classifying.

- Provide an outdoor environment with natural materials such as boulders and rocks.

- Light box exploration of natural materials can create a sense of awe and wonder.

- Provide young children with natural materials such as different coloured rocks, shells, flowers, leaves, shiny gems and a blank canvas. This allows for creativity and many learning possibilities about earth and biological science.

> *Children will become more deeply involved when you provide something that is new and unusual for them to explore, especially when it is linked to their interests.*
>
> Characteristics of Effective Learning, Active Learning, motivation, 2012, Development Matters

- Organise treasure hunts to look for special treasures, such as small rocks, leaves and other precious items. Provide treasure boxes so that children can store their collections.

- The outdoors provides many opportunities for young children to observe the colours of nature. A simple colour-matching experience with sample colour cards from any hardware store will allow children to further develop and explore the characteristics of their world.

- Provide relevant picture and factual books about rocks to support children's interests and further learning.

- Create dramatic play experiences using natural and recycled materials to support further investigation in a fun and exciting way.

These children from a school for those with hearing impairments enjoyed ochre painting on paving tiles. These tiles were then used to edge the garden. All the children helped with the planting, enjoyed working in the garden and connected with nature.

- Provide aesthetically pleasing play experiences by using glassware and different types of stones, gems and rocks.

- Support mathematical learning about measurement with the use of natural materials and recycled items.

Soils

Soil is made up of all kinds of things: broken down rocks, minerals, and organic materials such as decomposed plant matter.

When designing outdoor spaces, the inclusion of a digging area will allow for further exploration and investigation of earth materials. Add equipment to the digging area, such as shovels, spades, buckets and digging forks. For good scientific experiences, these should be smaller sized 'real' tools, not plastic or 'play' toys.

Mud

There is little more important in our physical world than earth and water, and they are especially intriguing substances when they interact. Mixing soil, water and a range of other natural materials has endless possibilities for early years development and learning. The breadth and depth of what these experiences offer young children is truly remarkable.

> *Mud can be the most delightful and enjoyable play experience in nature. It allows for creativity and imagination.*
>
> **Jan White, *Making a Mud Kitchen***

Practitioners should encourage young children to play freely with mud, without the fear of getting dirty. This fear can easily be overcome by providing appropriate clothing, such as overalls and boots.

Benefits of mud play

- Connects young children with nature.

- It is an imaginative and open-ended experience, promoting creativity.

- It provides a fun, sensory and soothing experience.

Experiences with mud

Mud play allows opportunities for young children to explore the properties and possibilities of mud, water and soil. Digging areas with access to water and equipment such as spades, buckets and wheelbarrows can encourage children to create and recreate all sorts of experiences.

International Mud Day is celebrated each year on the 29th of June. All around the world young children are encouraged to play with mud on this day.

Sand

Sand play can be a wonderful way to introduce young children to earth science.

Benefits of sand play

- Promotes creativity.

- Connects young children with nature.

- Promotes physical activity.

- Sandpits provide wonderful settings for social interaction.

- Playing with sand provides a different sensory experience than playing with mud (soil).

Experiences with sand

- Add props to sand play, such as funnels, sieves, cooking utensils, colanders, funnels, pots, cups, trucks and pipes.

- Include sand play indoors; add natural materials to support imaginative play and exploration with sand.

Day and night

Young children can usually make simple observations about day and night. Practitioners can help children to further develop their understanding about day and night.

- Discuss the differences between day and night. Ask open-ended questions, such as 'What kinds of things do you do during the day and/or night?'

- Provide opportunities for young children to record their knowledge and thoughts about day and night through drawing, painting and taking photos.

- Read relevant stories, such as Owl Babies by Martin Waddell.

- Provide factual books and posters relating to day and night.

- Invite children and families into your centre at night so children can observe the night sky and see what the centre looks like at night.

- Encourage young children to investigate their own shadows on sunny days.

- Take photographs of children and their shadows.

- During story time, include shadow puppet stories.

- Use overhead projectors to explore shadows indoors.

- Create simple yet inviting play spaces to stimulate children's learning about science.

Weather and seasons

Seasons in early years learning environments should focus on changes in the immediate environment.

- Discuss the seasonal changes in weather.

- Talk to children about seasonal fruit and vegetables and what they like to eat.

- Discuss the things children like to do and what they like to wear at certain times of the year; for example, do they like to go to the beach in summer?

- Talk to young children about the weather before going outside. Is it cold, windy, sunny, warm or hot? Draw their attention to the weather by simply looking out of the window and discussing what you see.

- On wet days, invite young children to jump in puddles and play with mud.

- On windy days, provide streamers, ribbons, balloons and bubbles. Children can observe these items flowing in the wind.

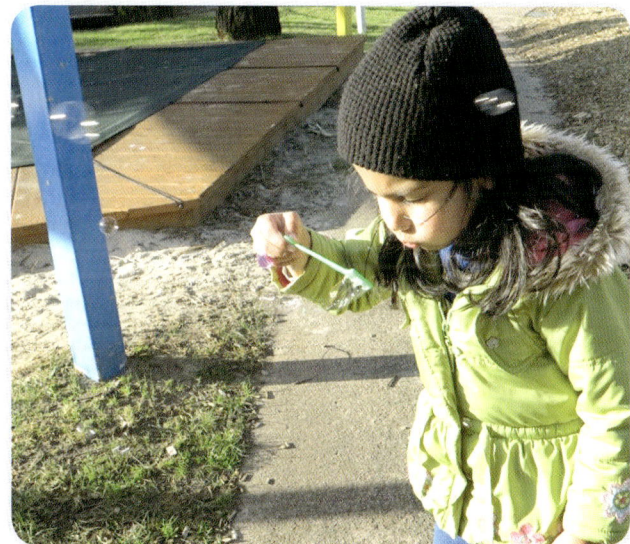

- Provide wind vanes and rain gauges within your setting so that children can observe changing weather.

- Display photos of seasonal changes in nature.

> *Give children time to talk and think. Value questions, talk, and many possible responses, without rushing toward answers too quickly.*
>
> Characteristics of Effective Learning,
> Creating and thinking critically,
> Development Matters, 2014

The seasons

- Autumn and winter are exciting seasons outdoors. The cooler temperatures in these seasons, along with rain, frost, snow and ice, make for wonderful and exciting times for young children.

- Help children appreciate the beauty and magic of the seasons by going for a walk with them in their immediate natural environment.

- Ask children to observe the changes they can see in nature. Draw their attention to growing flora and fauna and the colour changes in the leaves; suggest they collect and sort the leaves and other natural objects they find on the ground; encourage them to do observational drawings of things they see and collect from the outdoors.

Children's observational drawings of leaves

- Draw children's attention to the beauty of nature by planting ornamental plants in your setting.

- During spring allow young children ample opportunities to enjoy time outside, observe and notice the new growth all around.

- Encourage children to find the new growth on trees which had no leaves during the winter.

- Remind children to watch out for the arrival of birds – can they see any nests?

- Encourage children to express their observations, thoughts and ideas through an art collage.

> *Nature champions a beautiful perspective of the world.*
>
> Sandra Duncan

- Winter is a great time to get outdoors. Remember, there is no such thing as unsuitable weather so get your coats and boots, and get outdoors, where the children can observe the structure and physical appearance of trees without leaves.

- Young children can further investigate the natural environment with the use of scientific equipment such as cameras, binoculars and magnifying glasses.

- Draw children's attention to the weather conditions of winter, such as rain, wind, frost, hail, snow, lightning and thunder.

- Go for a walks in the rain.

- Children can help with watering the garden.

- Grow sunflowers in your outdoor garden.

> *Encourage children to try new activities and to judge risks for themselves. Be sure to support children's confidence with words and body language.*
>
> Characteristics of Effective Learning, Creating and thinking critically, Development Matters, 2014

- Go for a snail hunt or a caterpillar search.

- Summer is the ideal time for children to have fun and explore the outdoors. It is also an important time for children to learn about protecting themselves from the harmful elements of the sun and hot weather.

- Include water play in all seasons.

- Grow trees and plants that will attract butterflies and birds to your setting.

- Native plants and flowers will attract nectar feeding birds and pollinating insects.

- Create water features, such as a bird bath, to attract native bird life to your environment.

Chapter 8

Environmental science

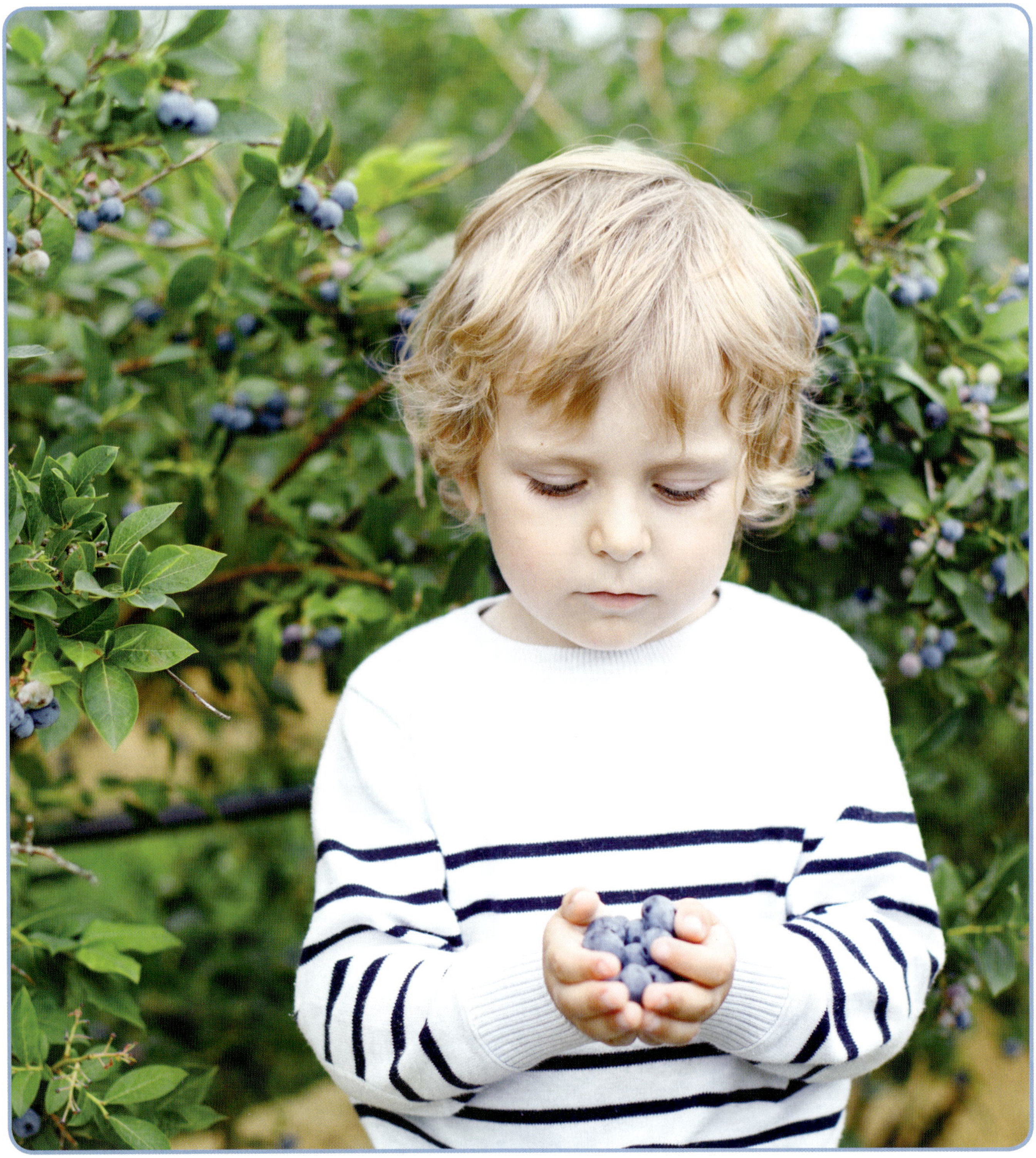

> *The world is not left to us by our parents. It is lent to us by our children.*
>
> African proverb

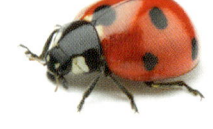

Environmental science in the early years

> *If we want children to flourish, to become truly empowered let us allow them to love the earth before we ask them to save it.*
>
> David Sobel

What is environmental science?

Environmental science in early years education is about understanding and caring for the natural world around us.

Why is it important?

Scientific learning is of little importance if we cannot respect, protect and sustain the environment in which we live. By exploring and participating in simple investigations within their natural world, young children will develop an awareness of and respect for the environment.

Promoting values, beliefs and thoughts about sustainability at an early age leads to the development of environmentally accountable adults.

Research shows that young children need to connect with nature for healthy growth, development and learning. They need time to wonder, explore, experiment, create and learn about their natural world. Children need play spaces that put them in touch with the beauty of nature, which will ignite their creativity, curiosity and imagination.

Richard Louv, in his important book, *The Last Child in the Woods*, suggests that nature is something to watch, appreciate, protect and consume. It is our role as early years practitioners to provide an environment where children can connect with nature.

Forest Schools

More than ever, there are many initiatives focused on using the outdoors as a total or partial learning environment.

- Forest Schools, originally a concept from Scandinavia, are now well–established in the UK.

  ★ England: www.forestschools.com

  ★ Northern Ireland: nifsa.org.uk

  ★ Wales: www.forestschoolwales.org.uk

  ★ Scotland: www.foresteducation.org

> " Children talk about how they and others show feelings, talk about their own and others' behaviour, and its consequences, and know that some behaviour is unacceptable. They work as part of a group or class, and understand and follow the rules. They adjust their behaviour to different situations, and take changes of routine in their stride.
>
> Early learning goal for Managing feelings and behaviour, EYFS, 2014 "

- The Forest School philosophy has been expressed in this way:

  At Forest School, all participants are viewed as:

  ★ Equal, unique and valuable

  ★ Competent to explore and discover

  ★ Entitled to experience appropriate risk and challenge

  ★ Entitled to choose, and to initiate and drive their own learning and development

  ★ Entitled to experience regular success

  ★ Entitled to develop positive relationships with other people

  ★ Entitled to develop a strong, positive relationship with their Forest School association (www.forestschoolassociation.org)

- Nature Kindergartens are also a familiar part of the education scene in Early Years Education. Auchlone Nature Kindergarten was one of the first.
  www.mindstretchers.co.uk/nature-kindergartens.efm

- Guelph outdoor pre-school declares that 'No child is left inside!'
  http://guelphoutdoorpreschool.com

- The Secret Garden Outdoor Nursery in Fife, Scotland, has no building – the children spend all day outside, for 49 weeks of the year.
  www.secretgardenoutdoor-nursery.co.uk

The practitioner's role

Practitioners should:

- Demonstrate a personal interest, wonder and appreciation of the natural world. It is important to remember that adults also need time to spend with nature for their own wellbeing.

- Celebrate the small events of the natural world – the spider web, the ant colony, the rainbow or the puddle. Even in a small outdoor area, you can explore nature and science if you look hard enough!

- Promote learning about respecting and caring for their natural world – for example, protecting wildlife, recycling and waste management.

- Encourage and engage young children in outdoor activities such as gardening.

- Provide natural materials that are rich in texture and colour.

> *If you are thinking a year ahead sow a seed;*
> *If you are thinking ten years ahead, plant a tree;*
> *If you are thinking a hundred years ahead,*
> *educate the people.*
>
> Chinese proverb

Some starting points for investigation

- The relationship between ourselves and the environment in which we live.

- Ways we can sustain, respect, protect and care for our world.

Scientific learning possibilities

- Developing an appreciation of the natural world.

- Beginning to understand about protecting the world – plants, animals and the land.

Developing science process skills

You can do this by asking questions relating to the natural environment, observing, problem-solving, classifying, and predicting.

> *Children learn and develop well in enabling environments, in which their experiences respond to their individual needs and there is a strong partnership between practitioners and parents and carers.*
>
> Development Matters, EYFS, 2014

Experiences for children from birth to 3 years

- Provide babies and younger children with natural and recycled materials for play experiences including shells, stones, cones and smooth pieces of wood.

- Go for walks outdoors. This will help children of all ages to experience the magic and wonder of their natural world.

- Encourage babies and toddlers to use their senses by seeing, touching, smelling, tasting and listening to objects in the environment.

- Provide opportunities for babies and toddlers to explore their natural environment by collecting objects such as leaves, feathers and other treasures.

- Read books and display photos of natural environments.

- Allow for quiet time and places to relax, within the natural environment.

- Ensure that all children have some natural surfaces to explore, e.g. natural grass for babies to crawl on.

- Help take care of plants in the outdoor and indoor environment by watering, pruning, dead-heading flowers and collecting leaves.

- Involve all children in caring for local wildlife.

- Attract birds into your centre with bird-friendly plants, nesting boxes, bird baths and hanging bird feeders.

- Encourage pollinating insects by planting flowers and shrubs.

Provide an aesthetically appealing environment for children by using equipment and props made from natural materials. For example, use wooden bowls with smooth textures and rich colours. Children can enjoy looking at and feeling these items, holding them, banging them together, and carrying them around the room.

Help the environment by using recycled and reused items such as mirrors, ribbons, lace, water bottles and plastic crates. Displaying these or offering them as resources will help to inspire exploration and creativity.

Keep animals at your setting. Chickens are environmentally friendly pets: toddlers and preschoolers can help to care for chickens by feeding, cleaning and maintaining their hen house and also by collecting eggs. See page 50 for more advice on keeping chickens.

Experiences for children from 4 to 5 years

- Provide a wide range of resources to support research and investigation, including posters, internet access, factual texts and story books.

- Continue to provide natural resources (shells, stones, corks, cones, bark, sticks, mud, sand and gravel) and 'loose parts' (guttering, drainpipes, string and rope, tape and fasteners, fabric pieces, tyres etc.) for free play.

- Establish a frog pond, a snail enclosure.

- Hatch chicks or butterflies.

- Grow native plants, flowers and vegetables in your outdoor area.

- Provide opportunities for children to climb and explore trees.

- Go walking in your local area, in parks, gardens, fields and playgrounds.

- Involve young children in gardening activities, such as digging up garden beds, and planting and harvesting vegetables.

- Share vegetables, herbs, fruits and flowers with families by having cooking, barbecue or picnic times with them.

- Involve children, staff and families in designing your vegetable garden.

- Don't use pesticides on the garden — get the children to help make a non-toxic home brew. using the following method.

Home brew to deter pests!

What you need:

1 litre of water

1 cup of oil

2 garlic cloves

A small piece of bath soap

Boil all ingredients together.

Cool and strain into a spray bottle.

- Participate in excursions to various places such as farms, local parks and botanical gardens. Zoos also offer excellent educational programmes and conservation campaigns.

- Visits to or visitors from organisations such as Farms for Schools (www.farmsforschools.org.uk), or Learning Through Landscapes (www.ltl.org.uk) who offer support for environmental and scientific learning.

In the outdoor area of this setting, an inviting play space has been created using recycled tyres.

Recycled pots and pans are included in sand play at this setting.

- Use natural materials and recycled equipment in creative ways in the outdoor environment.

- Create inviting play spaces with recycled tyres.

- Include children and families in discussions and decisions on waste management and water conservation.

- Introduce the philosophy of 'Rethink, Reduce, Reuse, Recycle' – for example, using scrap paper for paper-making, collage, drawing and painting; using recycled glass containers for paint; and reusing resources such as pre-loved furniture.

- Encourage young children to use recycled water from tanks for sand and mud play and for watering the garden.

- Get children involved in conducting an energy audit to find out how much energy your setting is using and wasting. Give children responsibilities, such as being energy monitors.

Some settings use 'pre-owned' furniture.

- Use overhead projectors with natural, recycled and reused materials, such as scrap paper, plastic bottles, feathers and leaves. This will allow young children to explore items in detail with the use of light. Children are also using their technology skills to switch the overhead projector on and off.

- Ask families and local businesses for usable and safe items, such as paper, wood, pots, glassware, pipes, hoses and furniture for children to use

- Visit or join a Scrapstore Project (www.scrapstoresuk.org) to find recycled and waste materials to use for science or creative activities.

The scarecrow at this setting is made from recycled plastic materials.

- Encourage young children to reuse items, such as cardboard boxes, paper, fabric, wool and plastic materials in creative play experiences.

- Provide recycling bins for children and parents to use.

The following pages include more detailed ideas for exploring environmental science with 4- to 5-year-olds.

Composting

What is composting?

Compost is formed when organic matter – for example leaves, grass clippings, paper and kitchen scraps – is changed by decomposition into rich and beautiful garden fertiliser.

Compost bins are essential in an outdoor learning environment. Composting is a practical way for an early years setting to reduce its waste.

You will need:

- A compost bin (available at garden centres or some local councils) or you can build your own compost heap

- Vegetable and fruit scraps, weeds, grass clippings, paper, cardboard, shredded newspaper, dried leaves and garden clippings

- Watering cans

- Metal garden forks

Composting activities for children

- Collecting and adding scraps to the compost bin.

- Incorporating air by turning the compost heap regularly with a garden fork.

- Adding more materials such as dried leaves, if the compost becomes too wet.

- Adding water, so that the compost heap is kept moist.

- When scraps have decomposed, adding compost to garden beds, plants and fruit trees — they will love it!

Scientific learning possibilities

- Learning how to take care of the environment.

- Developing an awareness of healthy environments.

- Learning about the importance of recycling organic waste.

- Learning how to produce compost.

Developing science process skills

Observing closely, asking questions, investigating, hypothesising, problem-solving, predicting and interpreting.

Other areas of learning

- Developing physical fitness – turning, digging and moving compost.

- Language skills – communicating and discussing observations and findings.

- Social skills – working together as a team.

- Mathematical skills – sorting, classifying, measuring, tracking of time, estimating depth of compost bin.

Worm farming

Worm farming can be an interest for children and will benefit the garden by providing rich fertiliser for the soil. Worms can get through great amounts of organic waste and are wonderful helpers in reducing the amount of waste sent to landfill.

You will need:

- A worm farm container – available from local councils and garden suppliers

- Live compost worms, which are usually sold separately and are available from garden suppliers

- Organic matter, e.g. vegetable and fruit scraps, cardboard, newspapers, leaves, soil, straw, grass

- A shady spot (worms don't like light)

- Damp (not wet) newspaper torn into strips

Worm farm activities for children

- Stacking the layers of the worm farm on top of each other.

- Collecting, sorting and adding scraps to the upper layer of the worm farm.

- Keeping the worm farm moist (worms die in dry soil, but drown in water – worms cannot swim).

- Collecting worm castings (worm poo) every few weeks.

- Spreading worm castings onto the garden – it makes rich organic fertiliser.

- Collecting liquid from the tap at the base of the worm farm, then dilute by adding water to make liquid fertiliser.

- Watering vegetable garden and flowers with your fertiliser.

Scientific learning possibilities

- Developing an understanding that they can actively help in caring for their world.

- Learning about how to reduce the amount of organic waste sent to landfills.

- Closely observing and investigating soil.

- Observing how worms grow and multiply.

- Learning to care for, respect and handle worms.

- Learning to use organic fertiliser.

Developing science process skills

Observing, asking questions, problem solving investigating and interpreting.

Other areas of learning

- Mathematical skills – sorting food scraps.

- Language skills – discussing and communicating their findings.

- Persistence and patience – waiting for decomposition to occur and collecting worm castings and fertiliser every few weeks.

- Social and emotional development – working together and building self-confidence.

- Wellbeing – developing environmentally sound practices.

Worms like: fruit, vegetables, leftover cereal, bread, moist cardboard and paper, tissues, hair, used tea bags and coffee grounds

Worms don't like: citrus fruits, onions, dairy products, raw potato, oil and meat

For more information on worm farms, visit your local council or go to **www.bbc.co.uk/gardening/…/ homegrownprojects_watchworms.shtml** for information.

References, further reading and resources

Statutory curriculum documents

United Kingdom

DFE, March 2014, Effective September 2014, *Statutory Framework for the Early Years Foundation Stage*, Setting the standards for learning, development and care for children from birth to five, **www.gov.uk**

Early Education, 2012, *Development Matters in the Early Years Foundation Stage* (EYFS), **www.early-education.org.uk**

Wales

Welsh Assembly Government, 2008, *Framework for Children's Learning for 3 to 7-year-olds in Wales*, **http://learning.wales.gov.uk**

Welsh Assembly Government, 2008, *Personal and Social Development, Well-Being and Cultural Diversity*, **www.wales.gov.uk**

Scotland

Learning and Teaching Scotland, 2010, *Pre-Birth to Three, Positive Outcomes for Scotland's Children and Families*- **http://www.scotland.gov.uk**

Learning and Teaching Scotland, 2010, *Curriculum for Excellence Principles and Practice*, **www.educationscotland.gov.uk/thecurriculum**

Northern Ireland

CCEA (Northern Ireland), *Learning Through Play in the Early Years* **www.nicurriculum.org.uk**

CCEA, *The World Around Us*, from Curricular Guidance for Pre-school Education; Northern Ireland

CCEA, 2008, *Revised Northern Ireland Curriculum* (RNIC), Implementation of The World Around Us, **www.nicurriculum.org.uk**

Early Years Inter-board publication: The Council for the Curriculum, Examinations and Assessment (Northern Ireland), 2006, *Understanding the Foundation Stage*, **www.nicurriculum.org.uk**

The Council for the Curriculum, Examinations and Assessment (Northern Ireland), 2008, *Curricular Guidance for Pre-school Education*

Australia

Early Years Learning Framework - **https://education.gov.au/early-years-learning-framework**

References and further reading

Amsel, S. *The Everything Kids' Environment Book: Learn how you can help the environment-by getting involved at school, at home, or at play*. Adams Media. 2007

Arbor Day Foundation. *Learning with Nature Idea Book*. Arbor Day Foundation. 2008

Beeley, K. *Science in the Early Years Foundation Stage: Hundreds of Ideas for Science-based Learning in the Early Years*. Featherstone Education. 2012

Beeley, K. *50 Fantastic ideas for Science Outdoors*. Featherstone Education. 2014

Brunton, P. and Thornton, L. *Science all Around*, Early Education, **www.early-education.org.uk**

Brunton, P. and Thornton, L., *Science in the Early Years: Building Firm Foundations from Birth to Five*. Sage. London. 2011

Brunton, P. and Thornton, L. *Science All Around*. Early Education. **www.early-education.org.uk**

Carson, R. *The Sense of Wonder*. Harper & Row. New York. 1956

Charlesworth, R. & Lind, K. *Math and Science for Young Children*. Delmar Learning. Clifton Park NY. 2003

Curtis, D. & Carter, M. *Designs for Living and Learning: Transforming Early Childhood Environments*. Redleaf Press. St Paul MN. 2014

Davis, G. & Keller, D. *Exploring Science and Mathematics in a Child's World*. Pearson Educa¬tion, Upper Saddle River NJ. 2009

Davis, J. *Young Children and the Environment Early Education for Sustainability*, Cambridge University Press, Melbourne. 2010

BeBoo, Max. Science 3-6: *Laying the Foundations in the Early Years*, ASE, Downloadable from **www.nationalstemcentre.org.uk/elibrary/resource/5973/science-3-6-laying-the-foundations-in-the-early-years**

Deviney, J. Duncan, S. Rody, M. & Rosenberry, L. *Inspiring Spaces for Young Children*. Gry¬phon House, Silver Spring, MD. 2010

Durie, J. Nemet, R. & Cameron R. *A Practical Guide for Kids in the Garden*. Jamie Durie Publishing, Sydney. 2005

Edwards et al. *The Hundred Languages of Children*. Ablex Publishing, Westport CT. 1998

Elliot, S. *The Outdoor Playspace Naturally: For Children Birth to Five Years*, Pademelon Press. Sydney. 2008

Elliot, S. & Emmett, S., *Snails Live in Houses Too: Environmental Education for the Early Years*. RMIT. Melbourne. 1997

Epstein, A. *The Intentional Teacher*. National Association for the Education of Young Children, Washington, DC. **www.naeyc.org**

Farmery, C. *Teaching science 3–11*: the essential guide. London: Continuum. 2002

Farrow, S. *The Really Useful Science Book: A Framework of Knowledge for Primary Teachers*, 3rd edn. Routledge. London. 2006

Featherstone, S. *Treasure Baskets and Heuristic Play*. Featherstone Education. 2013

Ferrari, A. & Giacopini, E. *ReMida Day: Mutta Menti*. Reggio Children. Reggio. 2006

Friedl, A. & Koontz, TY. *Teaching Science to Children: An Inquiry Approach*, 6th edn, McGraw-Hill. New York. 2005

Greenman, J. Caring Spaces, *Learning Places: Children's Environments That Work*. Exchange Press, Redmond, WA. 2005

Ha, T. *Green Stuff for Kids: An A to Z Guide to What's Up with the Planet*. Melbourne University Press, Melbourne. 2009

Harlan, J D. & Rivkin, M. *Science Experiences for the Early Childhood Years: An integrated affective approach.* 9th edn, Pearson Education, Upper Saddle River NJ. 2008

Keeler, R. Natural Playspaces: *Creating Outdoor Play Environments for the Soul.* Exchange Press, Redmond, WA. 2008

Kolbe, U. *Rapunzel's Supermarket: All About Young Children and Their Art* (2nd edn). Peppi¬not Press, Byron Bay NSW. 2007

Lind, K. *Exploring Science in Early Childhood: A Developmental Approach.* 4th edn, Thomson Delmar Learning. Clifton Park NY. 2005

Louv, R. *The Last Child in the Woods: Saving Our Children From Nature-deficit Disorder.* Algonquin Books, Chapel Hill, NC. 2008

Martin, D. *Constructing Early Childhood Science.* Wadsworth, Belmont, CA. 2001

Myer, C. Children as Artists, Early Education. **www.early-education.org.uk**

Rogers, V. Where does science fit in the Early Years? Primary Science. **www.ase.org.uk/journals/primary-science** . 2012

Seefeldt, C. & Galper, A. *Active Experiences for Active Children:* Science, 3rd edn, Pearson Education, Upper Saddle River NJ. 2012

Topal, C. & Gandini, L. *Beautiful Stuff! Learning with Found Materials,* Wyatt Wade, Worcester, MA.1999

United Nations. *The Convention of the Rights of the Child.* UNICEF, New York. 1989

Warden, C. & Buchan, N. *Look, Look, Look, Again, Spring.* Mindstretchers. Auchterarder, Perthshire. 2007

Warden, C. *The Potential of a Puddle.* Mindstretchers, Auchterarder, Perthshire. 2005

Warden, C. *Nurture Through Nature,* Mindstretchers, Auchterarder, Perthshire. 2007

Warden, C. *Nature Kindergarten: An Exploration on Naturalistic Learning within Nature Kindergartens and Forest Schools.* Mindstretchers, Auchterarder, Perthshire 2010

Wellhousen, K. & Crowther I. *Creating Effective Learning Environments.* Delmar Learning, Clifton Park NY. 2004

Wilson, R. *Nature and Young Children: Encouraging Creative Play and Learning in Natural Environments.* Routledge, London. 2008

Women's Weekly. *Kids in the Garden.* ACP Publishing, Sydney. 2011

Worth, K. *Science in Early Childhood Classrooms: Content and Process.* SEED Papers **http://ecrp.uiuc.edu/beyond/seed/worth.html 2010**

Young, T. & Elliot, S. *Just Discover: Connecting Children and the Natural World.* Tertiary Press, Melbourne. 2003

Children's books

Recycling (environmental)

Bethel, E. *Michael Recycle*. Worthwhile Books, Sydney. 2009

Green, J. *Why Should I Recycle/ Protect Nature/Save Water/Save Energy/?* Barron's Educational Series, New York. 2005

James, S. *Dear Greenpeace*. Walker Books, London. 1991

Growing things

Back, C. *Broad Bean*. A&C Black, London. 1986

Base, G. *Uno's Garden*. Penguin, Melbourne. 2006

Butterworth, N. *Jasper's Beanstalk*. Hodder Children's. 2008

Carle, E. *The Tiny Seed*. Little Simon, New York. 2009

Carle. E. *The Tiny Seed*. Simon Spotlight. New York. 2015

Crew, N. *Jack and the Beanstalk*. Henry Holt & Company, New York. 2011

Danks, F & Schofield, J. *The Stick Book*. Frances Lincoln Ltd, London. 2012

Eastman, PD. *Are You My Mother?* HarperCollins, London. 2006

Ehlert, L. *Growing Vegetable Soup.* Houghton Mifflin, 2013

French, J. *Diary of a Wombat*. HarperCollins, London. 2006

French, V. *Oliver's Vegetables*. Hodder Children's. 1995

French, V. *Oliver's Fruit Salad*. Hodder Children's. 1998

French, V. *Oliver's Milkshake*. Hodder Children's. 2000

Pinkney, J B. *Max Found Two Sticks*. Aladdin Paperbacks, New York. 1997

Animals

Carle, E. *The Very Hungry Caterpillar*, Puffin Book, London. 1995

Carle, E. *Polar Bear, Polar Bear, What Do You Hear?* Puffin Books, London. 2003

Carle, E. *Brown Bear, Brown Bear, What Do You See?* Puffin Books, London. 1997

Baker, J. *The Story of Rosy Dock*. Random House, Sydney. 2002

Base, G. *The Waterhole. Penguin,* Melbourne. 2001

Hen, LR. *The Little Red Hen and the Grains of Wheat*. Mantra Lingua, London. 2006

Oliver, N. *Baby Bilby, Where Do You Sleep?* Lothian, Melbourne. 2001

Humans

Bullard, L. *Marvelous Me: Inside and Out*. Picture Window Books. 2002

Carle, E. *From Head to Toe.* Puffin Books. London. 1999

Fox, M. *Whoever you Are*. Harcourt, Boston. 2007

Nettleton, P. *Look, Listen, Taste, Touch and Smell*. Picture Window Books, North Mankato. 2006

Read, L. *All About Me, My Senses*. Franklin Watts, London. 2012

Roca, N. *The 5 Senses*. Barron's Educational Series. 2006

Hutchins, P. *Titch*. Red Fox. 1997

Silverstein, S. *The Giving Tree*. HarperCollins, New York. 1964

Seuss, Dr. *My Book About Me*. HarperCollins, London. 1999

Sweeney, J. *Me and My Amazing Body*. Random House, New York. 2000

Sweeney, J. *Me and my senses*. Random House, New York. 2004

Powlay, L. *Little Topic Book of Ourselves*. Featherstone Education, London. 2010

Physical sciences

Aliki. *Fossils Tell of Long Ago*. HarperCollins, New York. 1990

Allen, P. *Who Sank the Boat?* Puffin Books, London. 1988

Allen, P. *Mr. Archimede's Bath*. Puffin Books, London. 1994

Baker, J. *Where the Forest Meets the Sea*. Walker Books, London. 1998

Bradley, KB. *Forces Make Things Happen*. HarperCollins, New York. 2005

Bradley, KB. *Pop! A Book About Bubbles*. HarperCollins, New York. 2001

Bulla, CR. *What Makes a Shadow?* HarperCollins, New York. 1994

Branley, F. *What Makes Day and Night?* HarperCollins, New York. 2007

Branley, F. *What Makes a Magnet?* HarperCollins, New York. 1996

Carle, E. *Draw Me a Star*. Puffin Books, London. 1995

Carle, E. *Papa Please Get the Moon for Me*. Neugebaurer Press, Boston. 1998, Cobb, V. I Fall Down. HarperCollins, New York. 2004

Christian, P. *If You Find a Rock*. Voyager Books, Boston MA. 2008

Lionni, L. *Alexander and the Wind-up Mouse*. Anderson Press, London. 2013

Lionni, L. *On My Beach There Are Many Pebbles*. HarperCollins, New York. 1995

Llewellyn, C. *And Everyone Shouted Pull: A First Look at Forces and Motion*. Picture Window Books, Mankato MN. 2004

Moss, L. *Zin, Zin, Zin! A Violin*. Simon & Schuster, New York. 2003

Pfeffer, W. *Light is all Around Us*. HarperCollins, London. 2015

Rey, M. *Curious George Flies a Kite*. Turtleback Books, St Louis, MO. 1999

Rey, H A. & Rey, M. *Curious George Visits a Toy Store*. Houghton Mifflin, New York. 2002

Wells, R. *How Do You Lift a Lion?* Albert Whitman & Company, Morton Grove IL. 1996

Zoehfield, KW. *What is the World Made Of? All About Solids, Liquids and Gases*. HarperCollins, London. 2015

Weather

Branley, F. *Snow is Falling*. HarperCollins. 2000

Carle, E. *Little Cloud*. Puffin Books, London. 1998

Cobb, V. *I Get Wet*. HarperCollins, New York. 2002

Ehlert, L. *Planting a Rainbow*. Voyager Books, New York. 1992

Ehlert, L. *Red Leaf Yellow Leaf*. Harcourt, Orlando. 1991

Germein, K. *Big Rain Coming*. Houghton Mifflin, Boston MA. 2001

Hutchins, P. *The Wind Blew*. Turtleback Books, St Louis MO. 1999

Maestro, B. *Why Do Leaves Change Colour?* HarperCollins, London. 2015

McKee, D. *Elmer and the Rainbow*. Andersen Press, London. 2007

Pfeffer, W. *Wiggling Worms at Work*. HarperCollins, London. 2007

Readman, J. & Roberts, L H. *George Saves the World by Lunchtime*. Eden Project Books, London. 2006

Sendak, M. *Where the Wild Things Are*. Bodley Head, London. 2001

Seuss, Dr. *The Lorax*. Collins, London. 1972

Silverstein, S. *The Giving Tree*. HarperCollins, New York. 1992

Tomlinson, J. *The Owl Who Was Afraid of the Dark*. Egmont, London. 2002

Tonkin, R. *Leaf Litter*. HarperCollins, Sydney. 2006

Vaughan, M. *Wombat Stew*. Scholastic, Gosford NSW. 2000

Waddell, M. *Owl Babies*. Walker Books, London. 1994

Mud

Hughes, S. *Mudlarks, in Out and About*, Walker Books. 2005

Buchanan, S. *Mud Pie Annie*. Zonder Kids. 2001

Lyn Ray, M. & Stringer, L. *Mud*. Voyager Books. 1996

Munsch, R. *Mud Puddle*. Annick. 2008

Funke, C. & Meyer, K. *Princess Pigsty*. Chicken House, Frome. 2007

Impey, R. *Joe's Café*. Orchard Books. 1993

James, B. & Morin, P. *The Mud Family*. Oxford University Press, Oxford. 1994

Cooper, H. *Pumpkin Soup*. Picture Corgi Books, London. 1999

Stockham, J. *Stone Soup*. Child's Play. 2006

Useful websites and organisations

Health and safety in early years settings

Be Safe, Association for Science Education, (England and Wales) http://www.ase.org.uk/resources/health-and-safety-resources/health-and-safety-primary-science/

CLEAPSS http://www.cleapss.org.uk/primary/primary-resources/primary-sras

BeBooMax, Science 3-6: Laying the Foundations in the Early Years, ASE, Downloadable from, www.nationalstemcentre.org.uk/elibrary/resource/5973/science-3-6-laying-the-foundations-in-the-early-years

Useful information

BBC Learning www.bbc.co.uk/learning/subjects/environmental_studies.shtml

Change for life www.nhs.uk/change4life/Pages/change-for-life.aspx

Composting www.recyclenow.com/recycle/recycle-school/composting

City Farms www.farmgarden.org.uk/

Early Childhood Australia www.earlychildhoodaustralia

Forest Schools www.forestschoolassociation.org

Scotland's Forest Schools/nurseries, and outdoor learning – the research www.educationscotland.gov.uk/learningteachingandassessment/approaches/outdoorlearning/supportmaterials/resources/Research.asp

Gardening with Children www.gardeningwithchildren.co.uk/

Greenpeace www.greenpeace.org.uk

Health Education Link Service http://healthykids.org.uk/

Keeping Chickens http://chickenschool.co.uk www.hensforhire.co.uk/hens-for-hire.html

Keeping pets in schools www.rspca.org.uk or www.pdsa.org.uk/about-us/education/pets-in-schools

Keeping stick insects www.keepinginsects.com/stick-insect/care/

Kids Health http://kidshealth.org/kid

Kids in the Garden http://kidsinthegarden.co.uk

Learning Through Landscapes www.ltl.org.uk

Mindstretchers www.mindstretchers.co.uk

The Mud Centre – Recapturing childhood through authentic mud play www.communityplaythings.com/resources/articles/dramaticplay/mudcenter.html

Nature Action Collaborative for Children www.worldforumfoundation.org

Nature Conservancy www.nature.org

National Geographic www.nationalgeographic.com

Oxfam www.oxfam.org.uk

Play England www.playengland.org.uk

Recycling Reggio Emilia www.reggiochildren.it/atelier/remida

Recycling www.scrapstoresuk.org

Reduce, Re-use, Recycle www.reducereuserecycle.co.uk/greendirectory/kids_green_sites.php

The Garden Classroom http://nurturestore.co.uk/kids-gardening-activities

UNESCO – Teaching and learning for sustainable future www.unesco.org/education/tlsf

Worm farming www.bbc.co.uk/gardening/.../homegrownprojects_watchworms.shtml

Young Peoples' Trust for the Environment http://ypte.org.uk/